T0226361

Bioinformatics for Evolutionary Biologists

Bernhard Haubold · Angelika Börsch-Haubold

Bioinformatics for Evolutionary Biologists

A Problems Approach

 Springer

Bernhard Haubold
Department of Evolutionary Genetics
Max-Planck-Institute for Evolutionary
   Biology
Plön, Schleswig-Holstein
Germany

Angelika Börsch-Haubold
Plön, Schleswig-Holstein
Germany

ISBN 978-3-319-88424-0   ISBN 978-3-319-67395-0   (eBook)
https://doi.org/10.1007/978-3-319-67395-0

Printed on acid-free paper

This Springer imprint is published by Springer Nature
The registered company is Springer International Publishing AG
The registered company address is: Gewerbestrasse 11, 6330 Cham, Switzerland

Preface

Evolutionary biologists have two types of ancestors: naturalists such as Charles Darwin (1809–1892) and theoreticians such as Ronald A. Fisher (1890–1962). The intellectual descendants of these two scientists have traditionally formed quite separate tribes. However, the distinction between naturalists and theoreticians is rapidly fading these days: Many naturalists spend most of their time in front of computers analyzing their data, and quite a few theoreticians are starting to collect their own data. The reason for this coalescence between theory and experiment is that two hitherto expensive technologies have become so cheap, they are now essentially free: computing and sequencing. Computing became affordable in the early 1980s with the advent of the PC. More recently, next generation sequencing has allowed everyone to sequence the genomes of their favorite organisms. However, analyzing this data remains difficult.

The difficulties are twofold: conceptual, which method should I use, and practical, how do I carry out a certain computation. The aim of this book is to help the reader overcome both difficulties. We do this by posing a series of problems. These come in two forms, paper and pencil problems, and computer problems. Our choice of concepts is centered on the analysis of sequences in an evolutionary context. The aim here is to give the reader a look under the hood of the programs applied in the computer problems. The computer problems are solved in the same environment used for decades by scientists, the UNIX command line, also known as the shell. This is available on all three major desktop operating systems, Windows, Linux, and OS-X. Like any skill worth learning, using the shell takes practice. The computer problems are designed to give the reader plenty of opportunity for that.

In Chap. 1, we introduce the command line. After explaining how to get started, we deal with plain text files, which serve as input and output of most UNIX operations. Many of these operations are themselves text files containing commands to be executed on some input. Such command files are called scripts, and their treatment concludes Chap. 1.

In Chap. 2, the newly acquired UNIX skills are used to explore a central concept in Bioinformatics: sequence alignment. A sequence alignment represents an evolutionary hypothesis about which residues have a recent common ancestor. This is determined using optimal alignment methods that extract the best out of a very large number of possible alignments. However, this optimal approach consumes a lot of time and memory.

The computation of exact matches, the topic of Chap. 3, is less resource intensive than the computation of alignments. Taken by themselves, exact matches are also less useful than alignments, because exact matches cannot take into account mutations. Nevertheless, exact matching is central to many of the most popular methods for inexact matching. We begin with methods for exact matching in time proportional to the length of the sequence investigated. Then we concentrate on methods for exact matching in time independent of the text length. This is achieved by indexing the input sequence through the construction of suffix trees and suffix arrays.

In Chap. 4, we show how to combine alignment with exact matching to obtain very fast programs. The most famous example of these is BLAST, which is routinely used to find similarities between sequences. Up to now we have only looked at pairwise alignment. At the end of Chap. 4, we generalize this to multiple sequence alignment.

In Chap. 5, multiple sequence alignments are used to construct phylogenies. These are hypotheses about the evolution of a set of species. If we zoom in from evolution between species to evolution within a particular species, we enter the field of population genetics, the topic of Chap. 6. Here, we concentrate on modeling evolution by following the descent of a sample of genes back in time to their most recent common ancestor. These lines of descent form a tree known as the coalescent, the topic of much of modern population genetics.

We conclude in Chap. 7 by introducing two miscellaneous topics: statistics and relational databases. Both would deserve books in their own right, and we restrict ourselves to showing how they fit in with the UNIX command line.

Our course is sequence-centric, because sequence data permeates modern biology. In addition, these data have attracted a rich set of computer methods for data analysis and modeling. The sequences we analyze can be downloaded from the companion website for this book:

http://guanine.evolbio.mpg.de/problemsBook/

To these sequences, we apply generic tools provided by the UNIX environment, published bioinformatics software, and programs written for this course. The latter are designed to allow readers to analyze a particular computational method. The programs are also available from the companion site.

At the back of the book, we give complete solutions to all the problems. The solutions are an integral part of the course. We recommend you attempt each problem in the order in which they are posed. If you find a solution, compare it to ours. If you cannot find a solution, read ours and try again. If our solution is unclear or you have a better one, please drop us a line at problemsbook@evolbio.mpg.de

The tongue-in-cheek Algorithm 1 summarizes these recommendations.

Algorithm 1 Using the Solutions

1: **while** problem unsolved **do**
2:     solve problem
3:     read solution
4:     **if** solution unclear **or** your solution is better than ours **then**
5:         drop us a line
6:     **end if**
7: **end while**

This book has been in the works since 2003 when BH started teaching Bioinformatics at the University of Applied Sciences, Weihenstephan. We thank all the students who gave us feedback on this material as it evolved over the years. We would also like to thank a few individuals who contributed in more specific ways to the gestation of this book: Mike Travisano (University of Minnesota) gave us encouragement at a critical time. Nicola Gaedeke and Peter Pfaffelhuber (University of Freiburg) commented on an early draft, and our students Linda Krause, Xiangyi Li, Katharina Dannenberg, and Lina Urban read large parts of the manuscript in one of the many guises it has taken over the years. We are grateful to all of them.

Plön, Germany   Bernhard Haubold
July 2017   Angelika Börsch-Haubold

The original version of the book backmatter was revised: For detailed information please see Erratum. The erratum to this chapter is available at https://doi.org/10.1007/978-3-319-67395-0_9

Contents

Chapter 1
The UNIX Command Line

Almost all commercial software published today comes with lush graphical user interfaces that allow users to work and play by touching and mousing. This is great for things like deleting a file by dragging it into a trash can, renaming a file by clicking on its name, editing text by mouse selection, and so on. However, in modern biology data may consist of dozens of files containing millions of sequencing reads, which makes it routinely necessary to do things like check the three billion nucleotides of the human genome for the occurrence of a particular motif, or compute averages from thousands of expression values distributed across dozens of files. Such operations are hard to perform using click-driven programs. This is because graphical user interfaces are excellent for carrying out the tasks their creators deem important, such as deleting a file by dragging and dropping it into a virtual trash bin, moving a file by dragging and dropping it between virtual folders, or opening a file by double-clicking on it. However, graphical user interfaces lack the universality that makes learning about computers so fascinating. Computers are universal machines in the sense that they can perform any precisely specified operation. All that is necessary is an interface that lets the user communicate every possible operation, not just a finite set, however large it may be.

To illustrate the importance of being able to communicate an infinite number of possible operations, think of the communication system we all know best, our language. Take any sentence that comes to mind and search the World Wide Web with it. Unless you were quoting from memory, chances are, your sentence is unique. This is because we do not parrot sentences we have heard, but use rules to construct new ones. The rules leave us free to think about what we want to say while saying it. Moreover, the words we use have a curiously vague relationship to what they mean. If someone says: "John is my friend.", the word "friend" neither looks nor sounds like a friend. Nevertheless, we know immediately what that word signifies. Taking our cue from language, we expect all powerful communication systems to be characterized by a set of rules and an arbitrary mapping between words and their meaning. Communicating effectively with a computer is no different.

© Springer International Publishing AG 2017
B. Haubold and A. Börsch-Haubold, *Bioinformatics for Evolutionary Biologists*, https://doi.org/10.1007/978-3-319-67395-0_1

The UNIX command line, also known as the *shell*, is the de facto standard method for text-based, rather than graphics-based, computer communication. It has been around since the late 1960s and has proved flexible enough to adapt rather than go extinct like so many other programs over the years. In fact, it is available on all three major operating systems, and its behavior is governed by a standard, the POSIX standard. This means that once you have mastered the UNIX shell on one type of computer, you have mastered it on all. If you have never used it, now would be a good time to start by working through the chapters in this part. Even if you have used it before, we recommend you work through this material to make the most of the subsequent sections. For future reference, the Appendix contains a summary of commands and techniques for working on the command line.

1.1   Getting Started

This section is for everyone who has never used the UNIX command line, or shell, before. There are various versions of the shell to choose from, but on personal computers bash is the default. We explain how to create and destroy directories and files under the shell, list the contents of directories, access the history of past commands, and help with typing. Fluency in typing is particularly important in a text-based system like the shell, and we encourage readers to spend time on practicing the basic key combinations. The chapter closes with a description of the manual system. We assume you are sitting in front of a computer with an open terminal displaying a blinking cursor like this:

```
jdoe@unixbox:$ ▮
```

New Concepts

Name	Comment
*	wildcard to match any substring
autocompletion	makes typing easier
UNIX	operating system
command line	text-based interface to UNIX

New Programs

Name	Source	Help
cd	system	man cd
ls	system	man ls
man	system	man man
mkdir	system	man mkdir
rmdir	system	man rmdir
rm	system	man rm
touch	system	man touch

Problem 1 Create a directory for this course by typing

```
mkdir BiProblems
```

followed by the Enter key. List the contents of your current directory to make sure BiProblems has been created.

```
ls
```

Notice that we write the names of directories in upper case to distinguish them from file names, which we start in lower case. This is merely a convention, others prefer to use lower case throughout. However, UNIX is case sensitive, so BiProblems, biProblems, and biproblems would be three distinct names. Notice also that we visualize word boundaries by case changes. Again, this is only a convention, known as "camel case". Change into BiProblems

```
cd BiProblems
```

and list its contents

```
ls
```

It is empty. Create two more directories, TestDir1 and TestDir2, and use ls to check they exist.

Problem 2 To minimize typos, the command line supports autocompletion. Change again into TestDir1, but this time type only

```
cd T
```

followed by Tab. This completes the unambiguous part of the name, TestDir. To get the two possible completions, press Tab again. Type 1, once more followed by Tab, to ensure correct typing. This technique of mixing typing and tabbing is very effective when using the shell. But it does take some getting used to. Practice it by changing out of the current directory

```
cd ..
```

and into it again. What happens if you enter

```
cd
```

without the trailing dots?

Problem 3 Use rmdir to remove the test directories. Then practice creating and removing these directories a few times. What happens if you enter

```
rmdir TestDir*
```

Problem 4 Recreate the directory TestDir1 and change into it. Then create a file

```
touch testFile
```

and check it exists

```
ls
```

Remove the file

```
rm testFile
```

Recreate the file, then go to the parent directory. What happens if you now apply `rmdir` to `TestDir1`?

Problem 5 Recreate `TestDir1` and enter it. Create two test files, `testFile1` and `testFile2`. How would you remove both with one command?

Problem 6 File a is renamed b by

```
mv a b
```

Create file a,

```
touch a
```

then try renaming it. Can you guess what `mv` might stand for?

Problem 7 Commands are often repeated. To avoid repeated typing, the command line remembers a list of previous commands. You can walk this list up and down by using the arrow keys ↑ and ↓. Try this. What happens when you enter the command

```
history
```

Problem 8 By now you have probably noticed that the cursor cannot be positioned by clicking the mouse. This leaves the arrow keys as the navigation tools of first choice. However, the cursor is also responsive to more powerful key strokes; for example, when you press the `Ctrl` followed by a, while still keeping `Ctrl` pressed, the cursor jumps to the beginning of the line. We write this as

```
C-a
```

Similarly,

```
C-e
```

moves the cursor to the end of the line. Type

```
You cannot tune a mouse but you can tuna fish
```

and practice jumping to the beginning and the end of the line a few times. What happens if you enter this nonsense as a command?

Table 1.1 List of key combinations for navigating and editing the `bash` command line

Keystrokes	Explanation
`C-e`	Position cursor at end of line
`C-a`	Position cursor at beginning of line
`C-w`	Delete word to the left of cursor
`C-y`	Insert deleted text
`C-b`	Move cursor back to one position
`C-f`	Move cursor forward by one position
`C-d`	Delete character left of cursor
`M-b`	Jump back by one word
`M-f`	Jump forward by one word
`M-d`	Delete word to the right of cursor

Problem 9 Table 1.1 lists the most useful key combinations for navigating the `bash`. Apart from the combinations based on the `Ctrl` key, there are also combinations based on the so-called `Meta` key, `M`. By default this is mapped to `Esc`. It may also be mapped to `Alt`, which makes it to easier to reach.

Moving the cursor using key combinations is a bit awkward at first, but once you have mastered these shortcuts, using the command line becomes much easier. Experiment with each of the key combinations in Table 1.1. What happens if you keep pressing a combination, say `C-f`?

Problem 10 If you need help with a command, or would like to learn more about its options, access the corresponding section of the manual by typing, for example

```
man ls
```

Navigate the page with the arrow keys, and press q to quit. It is often useful to know which file in a directory was modified most recently. Read `man ls` to find out how files can be listed by modification time.

Problem 11 Find out more about how to navigate the man pages by again typing

```
man ls
```

and then activate the help function by pressing h. How would you look for the pattern `time` in a man page?

Problem 12 A very useful feature of the shell is that the output of a program can be used as input for another. For example in

```
ls | wc
```

the program `wc` reads as its input the output from `ls`. This construction is called a "pipe" or "pipeline". Can you interpret the result of your first pipeline? How would you count the number of files in your directory?

Problem 13 How would you find out more about pipelines under the `bash`?

Problem 14 The `bash` is a programming environment and can be used as a simple calculator. To add two numbers, type, for example,

```
((x=1+1)); echo $x
```

where `echo` prints the value of `x` to the screen. What happens if you leave out the double brackets?

Problem 15 If you like a more verbose output, enter

```
((x=1+1)); echo "The result is $x"
```

What happens if the double quotes are replaced by a single quotes?

Problem 16 Our simple calculation can also be expressed as

```
let x=1+1; echo $x
```

What happens if you leave out the `let`?

Problem 17 The `bash` can also multiply

```
let y=2*5; echo $y
```

and compute power of

```
let y=2**5; echo $y
```

What is 2^{10}?

Problem 18 Subtraction also works as expected

```
let y=10-2; echo $y
```

What is $10 - 20$ according to the `bash`?

Problem 19 Division is denoted by

```
let y=10/2; echo $y
```

What is $10/3$?

Problem 20 Floating point calculations on the command line can only be carried out using additional tools. One of these is the basic calculator, `bc`. Enter

```
bc -l
```

to start it, and to exit `quit`. In bc n^x is expressed as `n^x`. What is the number of distinct oligonucleotides of length 10? Can you guesstimate the result?

Problem 21 As usual for UNIX programs, the basic calculator can also be used as part of a pipeline:

```
echo 10/3 | bc -l
```

What happens if you drop the `-l` option?

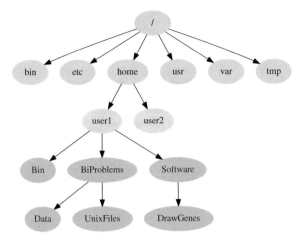

Fig. 1.1 The UNIX file system, slightly abridged. System directories are gray, home directories blue, and directories generated by users pink

1.2 Files

Files are kept inside directories, which may contain further directories. This hierarchy of directories forms a tree, and Fig. 1.1 shows an example containing the essential features of a typical UNIX file system. Its top node is the root, denoted /. The gray part of the tree is "given" and can only be changed by the system administrator, for example, when installing new software. The blue directories are called "home directories". Every user has one and can change it by adding new directories, which are depicted in pink. This separation between files accessible only to the administrator and files accessible to the user means that users need not worry about accidentally damaging the system—they cannot change the sensitive files. Our bioinformatics course forms a sub tree of the directory tree rooted on BiProblems, which you have already created. In the following section we learn to navigate the file system, and to manipulate individual files.

New Concepts	
Name	Comment
PATH	directories searched for file name
directories	contain files and directories
file permissions	read, write, execute
file system	all directories
files	contain text (usually)

New Programs

Name	Source	Help
apt	system	man apt
brew	system	man brew
cat	system	man cat
chmod	system	man chmod
cut	system	man cut
drawGenes	book website	drawGenes -h
emacs	package manager	man emacs
find	system	man find
gnuplot	system	man gnuplot
grep	system	man grep
head	system	man head
make	system	man make
tail	system	man tail
tar	system	man tar
which	system	man which

Problem 22   Use `ls` to list the contents of the root directory. How many files and directories does it contain?

Problem 23   Change into the course directory `BiProblems` and list all files it contains. Use `man ls` to find out how to really list *all* files. Can you explain what you see?

Problem 24   Download the example data from the book website

```
http://guanine.evolbio.mpg.de/problemsBook/
```

copy it into your current directory, and unpack it

```
cp ~/Downloads/data.tgz . tar -xvzf data.tgz
```

This generates the directory `Data`. How many files does it contain?

Problem 25   It is often convenient to list all files that contain a certain substring in their name. For example, all files with the extension `fasta`:

```
ls *.fasta
```

How many FASTA files are contained in the `Data` directory?

Problem 26   Make a directory for this session and change into it:

```
mkdir UnixFiles
cd UnixFiles
```

The file `mgGenes.txt` contains a list of all genes in the bacterium *Mycoplasma genitalium*. Copy `mgGenes.txt` from `Data` into the current directory. How many genes does *M. genitalium* have?

Problem 27 The command

```
cat mgGenes.txt
```

prints the contents of mgGenes.txt to the screen. What does cat -n do? Use it to re-count the entries in mgGenes.txt.

Problem 28 We often need to look at the beginning and the end of a file. This is done using the commands head and tail. Apply these to mgGenes.txt; can you make head or tail of what you see?

Problem 29 From our quick glance at the head and tail of mgGenes.txt, it looks as though genes at the beginning of the list are on the forward strand, genes at the end on the reverse. Since the list is ordered according to starting position rather than strand, this is intriguing. Do genes on the forward and reverse strand form separate blocks along the genome? To find out, use the program grep, which extracts lines from files matching a pattern, for example,

```
grep MG_12 mgGenes.txt
```

Use this and wc to count the genes on the forward strand and on the reverse strand. Do the counts add up to the total number of genes? If not, can you think of why?

Problem 30 To investigate whether one of the gene symbols contains the extra "-", we cut out the symbol column:

```
cut -f 5 mgGenes.txt
```

Which genes contain in their names "-", and which strand are they located on?

Problem 31 We are still trying to find out whether genes on the plus and minus strands form separate blocks along the *M. genitalium* genome. To exclude unexpected characters contained in the gene names, we cut out the first four columns of mgGenes.txt and extract the genes on the two strands. For subsequent analysis, we save the results by redirecting them to files using

```
grep pattern mgGenes.txt > pattern.txt
```

Save the genes on the forward strand in the file plus.txt and the genes on the reverse strand in the file minus.txt. Check again that the gene counts add up. Do the genes on the plus strand form a block along the genome (hint: use head -n)?

Problem 32 Redirection also works in reverse. Can you find out how to apply <?

Problem 33 Next, we need to use an editor. Our editor of choice is called emacs. If emacs is not installed on your system, please install it now. On many versions of Linux, including Ubuntu, this can be done using the cycle

```
sudo apt-get update          # update package database
apt-cache search emacs       # find suitable package
sudo apt-get install emacs   # install package
```

On OSX you might use

```
brew install emacs
```

if the homebrew package manager is installed. What does the `sudo` in the apt-commands above stand for?

Problem 34 To avoid the problem with the gene name containing a "-", we can open `mgGenes.txt` in `emacs` and remove the offending hyphen. Open `mgGenes.txt`:

```
emacs mgGenes.txt &
```

This opens a new window running `emacs`. What happens if you leave out the ampersand (`&`)?

Problem 35 Save `mgGenes.txt` to `mgGenes2.txt` and replace the "-" in `rpmG-2` and `polC-2` by underscore, "_". Use `diff` to check the differences between `mgGenes.txt` and `mgGenes2.txt`.

Problem 36 Use `head` and `tail` to directly extract lines 56 and 288 from files `mgGenes.txt` and `mgGenes2.txt`.

Problem 37 Many of the commands for navigating the `bash` listed in Table 1.1 have the same function in `emacs`. What are the exception(s)?

Problem 38 `emacs` is a powerful editor with a rather weak GUI. We recommend you take some time to learn the most important keyboard shortcuts, which are summarized in Table 1 of the Appendix. In addition, we recommend you work through the `emacs` tutorial, which is invoked by `C-h t`. What is the command for exiting `emacs`?

Problem 39 Apart from programs like `emacs`, which are supplied through public software repositories, there are a number of programs written specifically for this course. These are supplied as source files accessible through the book website. As a first example, download the program `drawGenes`. It is a good idea to keep source packages in the same place, so make a directory `Software` in your home directory and copy the source package of `drawGenes` into it. Then unpack it (c.f. Problem 24), change into the new directory and compile the code by typing `make`. This generates the program `drawGenes`. Again, programs are best kept in one place, so make the directory `~/Bin` and copy `drawGenes` into it. Return to your current directory. What happens when you try executing `drawGenes`?

Problem 40 To make the system aware of the new directory for executables, ˜/Bin, we need to change the set of directories in which the system looks for executables when it receives a command like ls. This set of directories is defined in the bash variable PATH. To alter PATH, open ˜/.bashrc in emacs and add the line

```
export PATH=~/Bin:$PATH
```

at the end. Then return to your current working directory and load the new PATH information

```
source ~/.bashrc
```

The first thing to do now is to test the old PATH is still working by trying to execute ls. If this fails, PATH needs to be reset. On Linux this is done by entering

```
source /etc/environment
```

on OSX by entering

```
source /etc/profile
```

Then try again to change PATH in .bashrc. Once this has worked, test that drawGenes is executable from within your working directory

```
drawGenes -h
```

This might seem like a long-winded method for installing programs. The good news is that .bashrc is always loaded when a new terminal is opened, so source only needs to be executed if .bashrc is changed during a terminal session. The next program installed manually just needs to be copied into ˜/Bin to become available to you everywhere. The command which locates an executable file; try for example

```
which drawGenes
```

Where is ls located on your system?

Problem 41 Apart from programs, we can also search for files using, for example

```
find ~/ -name "*.txt"
```

which looks for all files with the extension txt, starting in the home directory. How would you look up the location of .bashrc?

Problem 42 The program drawGenes converts gene coordinates like

```
100 400 +
600 1500 -
```

to figures like

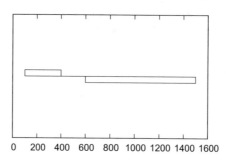

Create a new file called exampleGenes.txt in emacs and copy the gene coordinates. Then reproduce the above figure using drawGenes together with gnuplot. Hint: Check the usage of drawGenes by typing

```
drawGenes -h
```

Problem 43 The commands of gnuplot can be abbreviated to the first few unique characters. What is the shortest version of your gnuplot command for plotting the *M. genitalium* genes?

Problem 44 Gnuplot is a powerful tool with many options, which are summarized in a reference card posted on our book website. For example, the command

```
set xlabel "Label"
```

labels the x-axis. Use it to label the x-axis of our example plot with "Position (bp)".

Problem 45 Draw the genes of *M. genitalium*. Is the bias in their distribution between the strands visible?

Problem 46 When dealing with longer commands like the one for drawing the genes in *M. genitalium* (Problem 45), it is often more convenient to edit them in a separate file, which can then be executed by the bash. Such files are called "scripts". Copy the solution to Problem 45 into the file drawGenes.sh and run it

```
bash drawGenes.sh
```

What happens if you try to execute drawGenes.sh directly by typing

```
./drawGenes.sh
```

Problem 47 There are three kinds of file permissions: read, write, and execute. To inspect them, execute the *long* version of ls:

```
ls -l
total52
-rw-rw-r-- 1 haubold haubold 83      Mar 3 15:15 drawGenes.sh
-rw-rw-r-- 1 haubold haubold 13284 Mar 3 15:15 mgGenes.txt
-rw-rw-r-- 1 haubold haubold 13284 Mar 3 15:15 mgGenes2.txt
-rw-rw-r-- 1 haubold haubold 5157  Mar 3 15:15 minus.txt
-rw-rw-r-- 1 haubold haubold 6762  Mar 3 15:15 plus.txt
```

This shows the total size of the files in kilobytes, followed by information about individual files in eight columns:

1. File type and permissions: The first character is the file type; the two most important file types are ordinary file (–), and directory (d). The next nine characters are divided into three blocks of the three permissions already mentioned: read (r), write (w), and execute (x). Permissions not granted are shown as hyphens. The first three permissions concern the user, that is you, the second the group, and the third the world, which is everybody.
2. Number of links; for files this is usually one, but directories may contain a greater number of links.
3. User name.
4. Group name.
5. File size in characters.
6. Date on which the file was last altered.
7. Time when the file was last altered.
8. File name.

We can make `drawGenes.sh` executable:

```
chmod +x drawGenes.sh
```

Check the result

```
ls -l drawGenes.sh
-rwxrwxr-x 1 haubold haubold 83 Mar 3 15:15 drawGenes.sh
```

Now you can run

```
./drawGenes.sh
```

What happens if you drop `./` from this command?

Problem 48 To include scripts located in the current directory in the PATH, open `~/.bashrc` and change the line

```
export PATH=~/Bin:$PATH
```

to

```
export PATH=.:~/Bin:$PATH
```

How is the `bash` made aware of this change? Can you now directly execute `drawGenes.sh`?

1.3 Scripts

In Problem 46 we wrote our first script, `drawGenes.sh`, to help draw the 525 genes of *Mycoplasma genitalium*. Scripts are used extensively in bioinformatics. Throughout this book, we use three kinds of scripts: `bash`, `sed`, and AWK. `Bash` scripts are used to drive other programs. Sed scripts automate text editing, for example removing stray hyphens from gene names. Finally, AWK is a programming language for manipulating text files like mgGenes.txt. It is carefully described in a book by the authors of the language, Alfred Aho, Peter Weinberger, and Brian Kernighan, hence the name AWK [4].

New Concepts

Name	Comment
array	table in computer programs
hash	array indexed by strings
shell script	file containing commands
stream editor	in contrast to an interactive editor

New Programs

Name	Source	Help
awk	system	man awk
sed	system	man sed
seq	system	man seq
uniq	system	man uniq

1.3.1 Bash

Problem 49 Start this session by changing into the directory BiProblems. Then make a new directory, UnixScripts, and change into it. As we already saw in Problem 46, commands that work directly on the command line can usually be included in a bash script and then executed. The command we start off with is

```
echo Hello World!
```

Enter this on the command line. If we wanted to separate the two words by three blanks, we might try

```
echo Hello   World!
```

but this has the same effect as the original command. Try using single quotes to get the desired effect.

Problem 50 Scripts overcome the limitations of the command line as an editing environment. Write a script hello.sh containing

```
echo 'Hello World!'
```

A command can be repeated using a loop like

```
for((i=1; i<=10; i=i+1))   # i=1,2,...,10
do
    echo 'Hello World!'
done
```

where everything behind a hashtag is ignored and can be used for commenting. We can also write this script on a single line:

```
for((i=1; i<=10; i=i+1)); do echo 'Hello World!'; done
```

Run this code. What happens if you replace `echo` by `echo -n`?

Problem 51 An alternative way of looping in `bash` is

```
for i in $(seq 10)
do
    echo 'Hello World!'
done
```

Modify this loop such that it prints the numbers from 1 to 10 (hint: take a look at Problem 14).

Problem 52 Write the numbers from 1 to 10 on the same line.

Problem 53 We said that commands on the command line and in scripts are interchangeable. Execute

```
echo 5
```

on the command line. Find out by looking at the `man` page how to count in steps of two, or backwards.

Problem 54 We have already seen that the genes in *M. genitalium* are not distributed equally between the forward and the reverse strands along the genome. A simple way of visualizing this is to show the number of genes on one of the strands as a function of the number of genes surveyed. For this, copy first `mgGenes.txt` from `Data` to your current directory

```
cp ../Data/mgGenes.txt .
```

Then the command

```
cut -f 4 mgGenes.txt | head -n 100 | grep + | wc -l
```

counts the number of genes on the plus strand among the first 100 genes. Write a script that counts the number of genes on the plus strand among the first 1, 2, ..., 525 genes and save the script as `countGenes.sh`.

Problem 55 Edit your script such that it prints the number of genes on the plus strand as a function of the number of genes investigated. Then plot that function using `gnuplot`.

Problem 56 Loops in shell scripts can be nested. Edit `countGenes.sh` such that it prints the counts for the plus and the minus strands. Separate the two data sets by a blank line. Then plot the two functions in the same graph.

1.3.2 *Sed*

Problem 57 Instead of using an interactive editor like emacs to replace -2 by _2 in Problem 35, we could have used the stream editor sed:

```
sed 's/-2/_2/' mgGenes.txt
```

A construction like s/a/b/ is a small program: Substitute (s) some expression a by some expression b. Carry out the replacement of -2 by _2, and save the result in mgGenes3.txt. Use diff to check the new file is identical to your manual edit in mgGenes2.txt.

Problem 58 Next, we investigate how many genes have proper names. We start by cutting out the names in the fifth column, but still need to delete the blank lines:

```
cut -f 5 mgGenes.txt | sed '/^$/d'
```

where the sed command means, delete (d) a line whenever the start of a line (^) is followed directly by its end ($). How many of the 525 genes have a name rather than just an accession number?

Problem 59 Apart from substituting (s) and deleting (d), sed can print (p) particular lines, for example,

```
sed -n '56p' mgGenes3.txt
```

The option -n causes sed to *not* print non-matching lines. What happens if you leave out the -n? Find out by comparing the sed result to mgGenes3.txt using diff.

Problem 60 Sed can also output a range of lines:

```
sed -n 'x,yp'
```

where x is the starting line, y the end. Write a sed script that replaces head and check your result with diff.

Problem 61 In Problem 30 we used grep to find the gene names containing a hyphen ("-"). Use sed to carry out the same search.

Problem 62 What is the range of gene positions in *M. genitalium*? The entries in mgGenes3.txt happen to be sorted, and we could rely on that; but let us make sure and sort all start and end positions ourselves. First, we write all positions in a single column by replacing TAB by newline:

```
cut -f 2,3 mgGenes3.txt | sed 's/\t/\n/'
```

In case your version of sed does not allow this substitution, try the equivalent tr command

```
cut -f 2,3 mgGenes3.txt | tr '\t' '\n'
```

Next, sort the positions using `sort`. The default mode of `sort` is alphabetical. Find out how to sort the positions numerically to discover the smallest and the largest position.

Problem 63 What would happen if by accident you sorted the gene positions alphabetically?

Problem 64 Check that the genes in `mgGenes3.txt` are sorted by start position.

Problem 65 Next we ask, whether any of the genes in *M. genitalium* overlap. Here is a hypothetical pair of overlapping genes:

```
G1 1000 2000 +
G2 1990 3000 +
```

Does the genome of *M. genitalium* contain such configurations?

Problem 66 Several `sed` commands can be applied to the same input. For example, we might want to remove empty lines from the gene symbols *and* remove all underscores:

```
cut -f 5 mgGenes3.txt | sed '/^$/d;s/_//';
```

Instead of writing the `sed` commands on the command line, they can be written in a file, say `filter.sed`, and executed as

```
sed -f filter.sed
```

where `filter.sed` is

```
/^$/d  # delete empty lines
s/_//  # remove underscores
```

Gyrases are an important family of genes involved in the maintenance of DNA topology. How many gyrases does the genome of *M. genitalium* contain? User `filter.sed` in your answer.

1.3.3 AWK

Problem 67 A typical AWK program looks like this:

```
awk '{print $2}' mgGenes3.txt
```

Try out the code above; which column does it print? Print the last column.

Problem 68 It is not necessary to provide an input file. For example, enter

```
awk '{print "You entered: " $0}'
```

The program now waits for input and prints whatever is entered, which is referred to as $0. In AWK—like on the shell—two strings are concatenated simply by writing them next to each other. To exit, press C-c C-c. Write an AWK program that prints the sum of two numbers entered interactively by the user.

Problem 69 The general structure of an AWK program is

```
pattern {action}
pattern {action}
...
```

Without a pattern, all input lines are matched. A pattern might be

```
$4~/[+]/
```

to match lines where the fourth column contains a plus. Write an AWK program that prints only the genes on the plus strand; then write a variant that prints only the genes on the minus strand.

Problem 70 What happens if you leave out the action block in your previous command?

Problem 71 Recall that drawGenes converts input like

```
100 400 +
```

to a box above the zero line

```
100 0
100 1
400 1
400 0
```

and input like

```
600 1500 -
```

to a box below the zero line

```
600 0
600 -1
1500 -1
1500 0
```

which we can then plot as

Check this by comparing the output from

```
cut -f 2-4 mgGenes3.txt | head
```

to the output from

```
cut -f 2-4 mgGenes3.txt | drawGenes | head
```

Write an AWK program to carry out this transformation. Save it in drawGenes.awk and run it

```
awk -f drawGenes.awk mgGenes3.txt |
gnuplot -p pipe.gp
```

where pipe.gp contains

```
unset ytics
set xlabel "Position (bp)"
plot[][-10:10] "< cat " title "" with lines
```

Problem 72 Use AWK and sort to find the lengths and accession numbers of the longest and shortest genes in *M. genitalium.*

Problem 73 This program counts the lines in mgGenes3.txt:

```
awk '{c = c + 1}END{print "Lines: " c}' mgGenes3.txt
```

Notice the END pattern, which precedes a block executed once after all the lines in a file have been dealt with. A shorter way of expressing the line count is

```
awk '{c += 1}END{print "Lines: " c}' mgGenes3.txt
```

And since we are just adding 1 at every step, we can write even more succinctly

```
awk '{c++}END{print "Lines: " c}' mgGenes3.txt
```

Compute the average length of genes in *M. genitalium.*

Problem 74 We can save the lengths of all genes in the array len and then print them

```
{
      l = $3-$2+1
      len[n++] = l
}END {
      for(i=0; i<n; i++)
          print len[i]
}
```

An array can be thought of as a table with indexed entries; in the case of len the table looks like this:

index	value
0	1143
1	933
2	1953
...	
524	810

The variance of values x_i is defined as

$$s^2 = \frac{\sum_{i=1}^{n}(x_i - \overline{x})^2}{n - 1},$$

where $\overline{x}$ is their average. Modify the array-code above to determine the variance of gene lengths. Check your result using the program `var`, which is available from the book site. If there is a discrepancy between your result and `var`, try using `printf` instead of `print`:

```
printf "Var: %e\n", v
```

where `%e` is the "engineering" format used by `var`.

Problem 75 We have already seen that genes are not distributed uniformly between the forward and reverse strand along the *M. genitalium* genome and that the variance of their lengths is huge. Our next question is, are gene lengths also distributed nonuniformly along the *M. genitalium* genome? To investigate, again save the lengths in an array and then plot the cumulative length as a function of gene rank. Normalize the cumulative length such that it lies between 0 and 1 and plot it together with the value expected if gene lengths are distributed uniformly along the genome.

Problem 76 The program `uniq` finds unique entries in an alphabetically sorted list. Use `sort` and `uniq` to determine the number of unique gene names in mgGenes3.txt. Is any name repeated? Hint: Recall from Problem 58 how to remove empty lines.

Problem 77 Use `sort` and `uniq` to find the number of distinct gene lengths in *M. genitalium*.

Problem 78 The option `-c` switches `uniq` into counting mode. To find the most frequent gene lengths, numerically sort the output of `uniq -c`. What are the five most frequent gene lengths? Hint: Reverse-sort the output using the `-r` option of `sort`.

Problem 79 Here is an AWK version of `uniq -c`:

```
{
      count[$1]++
}END {
      for(a in count)
            print count[a], a
}
```

In contrast to `uniq`, this program works on unsorted and sorted input. Consider the construct `count[$1]++`. Since `$1` can be any string, not just a number, it is called a *key* rather than an index. And since consecutive index numbers are characteristic of arrays, `count` is called a *hash* instead of an array. The `for in` construct goes through all the keys of a hash. Notice also the comma in the `print` command for delineating two strings. Copy the AWK version of `uniq -c` into `uniqC.awk` and make sure it generates the same output as `uniq -c` (Problem 78). This is best done by removing the leading blanks in the output from `uniq -c` with

```
sed 's/ *//'
```

which means, substitute (`s`) one or more (`*`) blanks by nothing.

Problem 80 Use `uniqC.awk` to plot the count of gene lengths as a function of length.

Problem 81 To complement the `END` pattern, there is also a `BEGIN` pattern, which opens a block executed before any input lines are dealt with. This makes it possible to write "ordinary" programs, which are executed once. If, for example, we would like to generate a random DNA sequence in the program `ranSeq.awk`, we could write:

```
BEGIN{
    print ">RandomSequence"
    srand(seed)
    s[0] = "A"
    s[1] = "T"
    s[2] = "C"
    s[3] = "G"
    for(i=0; i<10; i++){
        j = int(rand() * 4)
        printf("%s", s[j])
    }
    printf("\n")
}
```

and execute

```
awk -v seed=$RANDOM -f ranSeq.awk
```

The output is in FASTA format, which means that each sequence gets a header line, which starts with >, followed by the sequence data on one or more lines. Notice the `-v` option, which allows variables in the program to be set on the command line. Write a version of `ranSeq.awk` which takes as input not only the seed for the random number generator, but also the sequence length. Note that the `-v` option needs to be repeated for every variable set on the command line.

Chapter 2
Constructing and Applying Optimal Alignments

2.1 Sequence Evolution and Alignment

DNA sequences evolve through mutations, insertions, and deletions of single nucleotides or small groups of nucleotides. We begin with a few paper and pencil exercises demonstrating the relationship between the evolutionary history of two DNA sequences and their alignment. This is followed by the computation of alignments.

New Concepts

Name	Comment
global alignment	homology across all residues
pairwise alignment	comparing homologous positions
sequence evolution	change over time

New Program

Name	Source	Help
gal	book website	gal -h

Problem 82 Consider a short example sequence, $S = $ ACCGT, which is passed from parent to child to grand-child, and so on. If replication were perfect, nothing would ever change. However, we only need to look at the biodiversity around us to remind ourselves that mutations do occur. Say, the G at position 4 in our example sequence changes into a C. Now the ancestral sequence has split into two versions, or alleles, which we can visualize as

© Springer International Publishing AG 2017
B. Haubold and A. Börsch-Haubold, *Bioinformatics for Evolutionary Biologists*, https://doi.org/10.1007/978-3-319-67395-0_2

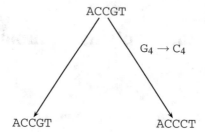

An alignment summarizes this scenario by writing nucleotides with a common ancestor on top of each other as follows:

ACCGT
ACCCT

Such nucleotides are called "homologous". Place a further mutation, an insertion, and a deletion along the lines of descent above. Write down the resultant sequences and their alignment.

Problem 83 With few exceptions, we can only sample contemporary sequences, while ancestral sequences remain unknown. Given two contemporary sequences, $S_1 = \text{ACCGT}$ and $S_2 = \text{ATGT}$, we wish to infer their evolutionary history by aligning them. One possible alignment is

ACCGT
ATGT-

The following is an evolutionary scenario compatible with that alignment:

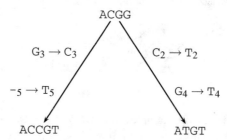

Draw an alternative evolutionary scenario leading to S_1 and S_2.

Problem 84 Consider again the two contemporary DNA sequences $S_1 = \text{ACCGT}$ and $S_2 = \text{ATGT}$ and write down five possible alignments. For each alignment note the minimal number of evolutionary events separating the two sequences since divergence from their hypothetical last common ancestor. A gap of length l is counted as l events. Here is an example:

ACCGT
ATGT-

There are three mismatches and one gap, hence four events.

Problem 85 To formalize the counting of evolutionary events, alignments are scored according to a score scheme, for example: match $= 1$, mismatch $= -3$, and gap,

$$g = g_0 + l \times g_e,$$

where $g_0 = -5$ denotes gap opening, l the gap length and $g_e = -2$ gap extension. Use this scheme to score your solutions to Problem 84.

Problem 86 Our gap score scheme implies that a newly opened gap is immediately extended by at least one step. How would you express the alternative view where gap opening itself leads to a gap?

Problem 87 Alignments are usually calculated with a computer. Go to the directory BiProblems and make the directory FirstAlignments. Change into it and print the example sequences $S_1 = $ ACCGT and $S_2 = $ ATGT in FASTA format (Problem 81) onto the command line. Find out what echo -e does and use it. What happens if you leave out the -e? Save the files in seq1.fasta and seq2.fasta.

Problem 88 Download gal from the course website and install it as explained in Problem 39. Check the usage of gal by typing

```
gal -h
```

Then use gal to align S_1 and S_2. Which gap scoring scheme is implemented by gal? Can you construct an alternative alignment with the same score?

Problem 89 Instead of playing with toy sequences, we now align two real sequences contained in hbb1.fasta and hbb2.fasta. Copy these files from Data to your current directory. What do these sequences encode? Align them using gal. Where do they differ?

Problem 90 Find the position of the mismatch; use the -l option of gal to make this easier.

Problem 91 Recall that hbb2.fasta is a partial CDS, which means it can be translated in frame starting at position 1. Does the single nucleotide difference between seq1.fasta and seq2.fasta lead to an amino acid change?

2.2 Amino Acid Substitution Matrices

DNA sequences are usually scored using a simple scheme involving only matches, mismatches, and gaps. However, pairs of amino acids all get their own score, which is summarized in substitution matrices such as the one shown in Fig. 2.1. There are two reasons for this: The structure of the genetic code and the diverse chemistry of

	A	R	N	D	C	Q	E	G	H	I	L	K	M	F	P	S	T	W	Y	V
A	5	-4	-2	-1	-4	-2	-1	0	-4	-2	-4	-4	-3	-6	0	1	1	-9	-5	-1
R	-4	8	-3	-6	-5	0	-5	-6	0	-3	-6	2	-2	-7	-2	-1	-4	0	-7	-5
N	-2	-3	6	3	-7	-1	0	-1	1	-3	-5	0	-5	-6	-3	1	0	-6	-3	-5
D	-1	-6	3	6	-9	0	3	-1	-1	-5	-8	-2	-7	-10	-4	-1	-2	-10	-7	-5
C	-4	-5	-7	-9	9	-9	-9	-6	-5	-4	-10	-9	-9	-8	-5	-1	-5	-11	-2	-4
Q	-2	0	-1	0	-9	7	2	-4	2	-5	-3	-1	-2	-9	-1	-3	-3	-8	-8	-4
E	-1	-5	0	3	-9	2	6	-2	-2	-4	-6	-2	-4	-9	-3	-2	-3	-11	-6	-4
G	0	-6	-1	-1	-6	-4	-2	6	-6	-6	-7	-5	-6	-7	-3	0	-3	-10	-9	-3
H	-4	0	1	-1	-5	2	-2	-6	8	-6	-4	-3	-6	-4	-2	-3	-4	-5	-1	-4
I	-2	-3	-3	-5	-4	-5	-4	-6	-6	7	1	-4	1	0	-5	-4	-1	-9	-4	3
L	-4	-6	-5	-8	-10	-3	-6	-7	-4	1	6	-5	2	-1	-5	-6	-4	-4	-4	0
K	-4	2	0	-2	-9	-1	-2	-5	-3	-4	-5	6	0	-9	-4	-2	-1	-7	-7	-6
M	-3	-2	-5	-7	-9	-2	-4	-6	-6	1	2	0	10	-2	-5	-3	-2	-8	-7	0
F	-6	-7	-6	-10	-8	-9	-9	-7	-4	0	-1	-9	-2	8	-7	-4	-6	-2	4	-5
P	0	-2	-3	-4	-5	-1	-3	-3	-2	-5	-5	-4	-5	-7	7	0	-2	-9	-9	-3
S	1	-1	1	-1	-1	-3	-2	0	-3	-4	-6	-2	-3	-4	0	5	2	-3	-5	-3
T	1	-4	0	-2	-5	-3	-3	-4	-1	-4	-1	-2	-6	-2	2	2	6	-8	-4	-1
W	-9	0	-6	-10	-11	-8	-11	-10	-5	-9	-4	-7	-8	-2	-9	-3	-8	13	-3	-10
Y	-5	-7	-3	-7	-2	-8	-6	-9	-1	-4	-4	-7	-7	4	-9	-5	-4	-3	9	-5
V	-1	-5	-5	-5	-4	-4	-4	-3	-4	3	0	-6	0	-5	-3	-3	-1	-10	-5	6

Fig. 2.1 PAM70 amino acid score matrix; match scores are shown in red

the encoded amino acids. According to the code, pairs of amino acids are separated by one, two, or three mutations. As to chemical diversity, the canonical amino acids also vary with respect to shape, polarity, charge, and hydropathy. In this chapter we explore how the genetic code and the chemistry of the encoded amino acids are incorporated into matrices such as Fig. 2.1 for scoring protein sequence alignments.

New Concepts

Name	Comment
conservation of pairs of amino acids	amino acids differ in evol. rate
matrix multiplication	simulate evolution
robustness of genetic code	has evolved

New Programs

Name	Source	Help
genCode	book website	genCode -h
histogram	book website	histogram -h
pamLog	book website	pamLog -h
pamNormalize	book website	pamNormalize -h
pamPower	book website	pamPower -h

2.2.1   Genetic Code

Problem 92 Our exploration of the genetic code follows a classic publication from the early 1990s [22]. Figure 2.2 shows the genetic code. There are $4^3 = 64$ codons

	T	C	A	G	
T	Phe/F	Ser/S	Tyr/Y	Cys/C	T
	Phe/F	Ser/S	Tyr/Y	Cys/C	C
	Leu/L	Ser/S	Ter/*	Ter/*	A
	Leu/L	Ser/S	Ter/*	Trp/W	G
C	Leu/L	Pro/P	His/H	Arg/R	T
	Leu/L	Pro/P	His/H	Arg/R	C
	Leu/L	Pro/P	Gln/Q	Arg/R	A
	Leu/L	Pro/P	Gln/Q	Arg/R	G
A	Ile/I	Thr/T	Asn/N	Ser/S	T
	Ile/I	Thr/T	Asn/N	Ser/S	C
	Ile/I	Thr/T	Lys/K	Arg/R	A
	Met/M	Thr/T	Lys/K	Arg/R	G
G	Val/V	Ala/A	Asp/D	Gly/G	T
	Val/V	Ala/A	Asp/D	Gly/G	C
	Val/V	Ala/A	Glu/E	Gly/G	A
	Val/V	Ala/A	Glu/E	Gly/G	G

4 5 6 7 8 9 10 11 12 13 14

Fig. 2.2 The genetic code with three-letter and single-letter amino acid designations, and color coding according to amino acid polarity

in total, of which three are stop codons. How long is the average open reading frame that starts with a start codon and ends with a stop codon?

Problem 93 The program `simOrf.awk` prints the lengths of open reading frames, ORFs in random DNA sequences. It can be run as

```
awk -v seed=$RANDOM -v n=10000 -f simOrf.awk
```

where `seed` is the seed for the random number generator and `n` the number of iterations. Test the predicted ORF length from Problem 92 using `simOrf.awk`.

Problem 94 Use `histogram` and `gnuplot` to plot the distribution of 1000 random ORF lengths.

Problem 95 The amino acids in Fig. 2.2 are color coded according to polarity. What are the two most polar amino acids?

Problem 96 How many mutations are necessary to get from phenylalanine (`Phe`) to leucine (`Leu`)? From `Phe` to tryptophane (`Trp`)? From `Phe` to glutamate (`Glu`)?

Problem 97 Figure 2.3 shows the side chains of all 20 amino acids. Among the many respects in which they differ, polarity is the most important [22]. For example, the

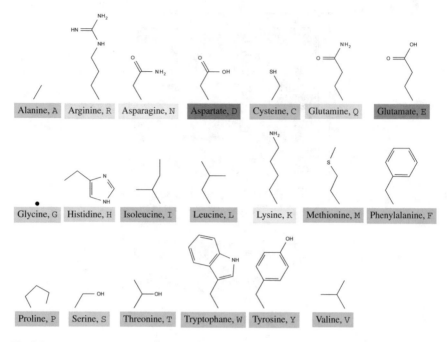

Fig. 2.3 The side chains of the 20 amino acids specified by the genetic code again color-coded by polarity. Glycine is merely bound to a hydrogen atom, the dot

aliphatic side chain of leucine has a much lower polarity than the side chain of glutamate, which is negatively charged at physiological pH. The file `polarity.dat` contains polarity values for all amino acids. Make the directory `AminoAcidMat` for this section, change into it and copy `polarity.dat` from `Data`. What is the most polar amino acid? The least polar?

Problem 98   The genetic code maps 64 codons to 20 amino acids. What is the largest number of codons encoding the same amino acid? The smallest?

Problem 99   A single nucleotide change in a codon can either leave the amino acid unchanged or not. Mutations that do not affect the encoded amino acid are called synonymous, non-synonymous their opposite. It is well known that mutations at the third codon position are often synonymous. Are there any synonymous mutations at the first two positions?

Problem 100   How many different genetic codes are there, if we leave the arrangement of codons unchanged and simply shuffle the 20 amino acids among them? If you are unsure, consider how many different arrangements there are for two books on a shelf, 3, 4, and so on.

Problem 101   Figure 2.4 shows the polarities of the 20 amino acids specified by the genetic code. Let us focus on phenylalanine, F, which is encoded by TT[TC].

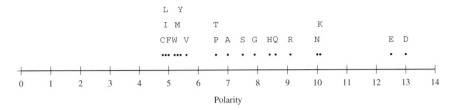

Fig. 2.4 Amino acid polarity, taken from [22]

Through the three possible single base mutations at the first position it can be turned into leucine (L), isoleucine (I), or valine (V). The polarity values of these four mutant amino acids are close to that of F (Fig. 2.4). Which amino acids can be reached through single base mutations at the second and third positions?

Problem 102 It seems as if there is a correlation between distances in codon space and distances in polarity space. To further explore this impression, the mean squared change in polarity is computed over all single base changes at every codon of the natural code. This quantity is called MS_0. Then the amino acids are shuffled between the codons and MS_0 is computed again. This computation is carried out by genCode. If you run

```
genCode polarity.dat
```

The program carries out the shuffling 10^4 times. If it finds a better code than the natural code, it prints it out. At the end of the run there is a little table of MS_0 values like

```
        ms0
nc      5.19
m(rc)   9.39
```

where nc refers to the natural code and m(rc) to the mean of random codes. The -p option prints all MS_0 values, not just those better than the $MS_0 = 5.19$ of the natural code. Locate 5.19 in that distribution. Is the natural code typical among the random codes?

Problem 103 In default mode genCode prints a random code if its MS_0 is less than that of the natural code. Run the program a few times with default parameters. What is roughly the proportion of random codes that are more polarity-robust than the natural code?

Problem 104 We can rephrase our search for better random codes as a null hypothesis: "The natural code is not mutation-optimized". What is the error probability when rejecting this null hypothesis? Use genCode with more than the default number of iterations to arrive at a robust answer.

Problem 105 Apart from polarity, amino acids differ also according to hydropathy, molecular volume, and isoelectric point. These quantities are stored in the files hydropathy.dat, volume.dat, and charge.dat. Is the genetic code optimized with respect to them, too?

2.2.2 PAM Matrices

The PAM matrices are the oldest amino acid substitution matrices still in use today. PAM stands for Percent Accepted Mutations [14]. This is the number of mutations per one hundred amino acids, as opposed to the percent difference between two amino acid sequences. The percent difference cannot grow beyond 100%, while the number of mutations that hit a certain stretch of sequence has no upper bound. A different way to think about PAM, is as a unit of evolutionary time: the time until one percent of the amino acids in a sequence have mutated, which is roughly three million years [24, p. 19]. This time dimension also suggests why we need different substitution matrices: In the limit of 0 PAM, all homologous amino acids are identical. So any mismatch would indicate a nonhomologous match and should carry a very low score. As time passes, the probability increases that two different amino acids are in fact homologous. Accordingly, a mismatch should carry a greater score than at PAM 0. To see how this insight leads to substitution matrices, we compute PAM matrices from scratch.

Problem 106 For the subsequent computations change into BiProblems, create the new directory PamComputation and change into it. Copy pam1.txt from Data into your working directory. Look at pam1.txt by typing

```
cat pam1.txt
```

It contains the mutation probabilities for all 20 amino acids after 1 PAM has elapsed. The entry M_{ij} indicates the probability that the amino acid in column j has mutated into the amino acid in row i. What is the mutation probability of alanine to serine? Serine to alanine?

Problem 107 The main diagonal of pam1.txt contains the probabilities of no change. For example, the probability that an alanine has not changed after 1 PAM is 0.9867. Next, we investigate the probability of drawing any two identical amino acids. For this we need to know for each amino acid the probability of finding it in a sequence. Let us pretend for now all amino acids have the same frequency, so the probability of finding a particular one is $1/20$. Then the probability of finding a pair of alanines is $1/20 \times 0.9867$, the probability of finding a pair of arginines $1/20 \times 0.9913$, and so on. What is the percent difference between homologous protein sequences after 1 PAM has elapsed?

Problem 108 Use the program pamPower to multiply pam1.txt n times with itself to generate M^n. This matrix multiplication simulates protein evolution for n

PAM units of time. Do this for $n = 1, 10, 100, 1000$. In the case of M^{1000}, what do you notice as you read along the rows of the matrix, compared to M^{10} and M^{100}?

Problem 109 Compute M^n for $n = 1, 2, 5, 10, 20, 50, 100, 200, 500, 1000$ and compute the percent difference between homologous amino acids each time. Plot this percent difference as a function of PAM.

Problem 110 Copy the file aa.txt into your working directory and print it to screen (cat). It contains amino acid frequencies in the same order in which the amino acids are listed in the fist row. What is the most frequent amino acid? The least frequent?

Problem 111 Compare the columns of $\mathbf{M}^{1000}$ with the amino acid frequencies in aa.txt. What do you observe?

Problem 112 Let us calculate a particular PAM matrix, say PAM70. We first compute the corresponding probability matrix using pamPower. The output of pamPower then needs to be "normalized" through division by the amino acid frequencies in aa.txt. This is done using the output of pamPower as the input to pamNormalize.

Problem 113 Use pamLog to calculate the final PAM70 matrix. Save this in pam70sm.txt (the sm stands for "substitution matrix"). Then calculate by hand the score of the following alignment using your PAM70:

ATLSE
SNLSD

Problem 114 Extract by hand all mismatched pairs of amino acids in PAM70 that have a score greater than zero. Look up the side chains of these amino acids in Fig. 2.3. What do you notice?

Problem 115 Use your PAM70 matrix together with gal to revisit Problem 89, where we aligned the RNA-sequences encoding hemoglobin stored in hbb1.fasta and hbb2.fasta on the DNA level. Now we align it on the protein level. Use transeq to translate hbb1.fasta and hbb2.fasta in all three forward reading frames to identify the correct frame. Like all EMBOSS-tools, transeq can read from stdin when executed with the -filter flag:

```
transeq -filter < hbb1.fasta
```

Save the resulting protein sequences in hbb1prot.fasta and hbb2prot.fasta, and align them. Can you spot the non-synonymous mutation?

Problem 116 What happens to the conservation of pairs of amino acids if we let the evolutionary distance between two protein sequences go toward infinity by computing PAM1000, PAM2000, and PAM3000? What is the most conserved amino acid?

Problem 117 What happens to the alignment of `hbb1prot.fasta` and `hbb2prot.fasta` if you use `pam1000sm.txt`, `pam2000sm.txt`, or `pam3000sm.txt`?

Problem 118 In Problem 107 we used 1/20 to approximate the amino acid frequencies. The program `percentDiff.awk` incorporates the exact amino acid frequencies in `aa.txt`. Run it like

```
pamPower -n 70 pam1.txt | tail -n +2 | awk -f
    percentDiff.awk
```

In Problem 109 we iterated the approximate %-difference computation using the script `pamPower.sh`. Copy the original script to `pamPower2.sh` and extend it to compare the two results. Plot both sets of results in one graph.

2.3   The Number of Possible Alignments

When looking for the best alignment, we might be tempted to construct all alignments and pick the one with the highest score. But before doing that, let us compute the number of alignments that can be formed between two sequences. Our calculation starts from the fact that every alignment ends in one of three ways as follows:

$$\boxed{\begin{array}{c} R \\ R \end{array}} \; , \; \boxed{\begin{array}{c} - \\ R \end{array}} \; , \; \text{or} \; \boxed{\begin{array}{c} R \\ - \end{array}}$$

where R stands for *residue* and might be an amino acid or a nucleotide. The first end implies that the remainders of both sequences to be aligned are one residue shorter; the other two ends imply that the remainder of one of the two sequences is one residue shorter. Hence, we can write the number of possible alignments between two sequences of lengths m and n as

$$f(m, n) = f(m - 1, n - 1) + f(m - 1, n) + f(m, n - 1). \tag{2.1}$$

In this function, f is a function of itself. This type of function is called *recursive*. As it stands, the recursion could go on for ever; but it is clear that sequences cannot have lengths less than zero. Moreover, if either of the two sequences (or both) have length zero, there is only one way to align them, for example

```
AATG
----
```

No other arrangement is possible, as columns of gaps are not allowed. We can summarize these observations as a set of three equations:

$$f(x, 0) = f(0, y) = f(0, 0) = 1, \tag{2.2}$$

which are called *boundary conditions*. In this section, we investigate two different approaches—one slow, the other fast—to evaluate Eq. (2.1).

New Concepts

Name	Comment
recursive function	a function of itself
top-down solution	"naïve" solution of recursive function
bottom-up solution	better solution of recursive function

New Program

Name	Source	Help
numAl	book website	numAl -h

Problem 119 Compute the number of possible alignments between two sequences of lengths two and three by directly solving Eq. (2.1). Draw a tree in which each term on the left of that equation is linked to the three terms on its right as follows:

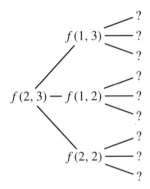

Problem 120 This direct approach to solving Eq. (2.1) is also called the *top-down* solution, as it proceeds from the most complex, "top", component of the problem down to its less complex parts. This leads to repeated computation of component terms. To avoid this, we can work our way up from the boundary condition. For this *bottom-up* solution, we need a matrix containing each possible term $f(i, j)$. For the example sequences of lengths 2 and 3 this is

	0	1	2	3
0				
1				
2				

where every entry $f(i, j)$ is the number of possible alignments between two sequences of lengths i and j. Compute $f(3, 2)$, by filling in this matrix, starting with the boundary conditions in Eq. (2.2) and then applying Eq. (2.1). For example, $f(1, 1)$ is found by adding its three neighboring entries, $f(1, 1) = f(0, 1) + f(0, 0) + f(1, 0)$.

Problem 121 Write down all possible global alignments between the sequences $S_1 = \text{AGT}$ and $S_2 = \text{AC}$.

Problem 122 Use the program numAl to compute the number of possible global alignments between two sequences of lengths 106.

Problem 123 Create the directory NumberOfAlignments and change into it. Write a shell script, numAl.sh, that drives numAl to compute the number of alignments between sequence pairs with equal lengths 1, 2, ..., 106. Plot the number of alignments as a function of sequence length. Use the command

```
set logscale y
```

in your gnuplot script.

Problem 124 Save numAl.sh to numAl2.sh to compute the number of possible alignments using the top-down solution. What do you observe? Hint: Remember that computations can be stopped using C-c C-c.

Problem 125 Plot the run time of the top-down solution with lengths 1, ..., 14. To make the log-transformation in gnuplot possible, filter out all zero run times.

Problem 126 Determine the linear function describing the graph you drew in Problem 125 and use this to calculate the number of years necessary to determine the number of possible alignments between two sequences of length 106 using the top-down solution.

2.4 Dot Plots

Dot plots provide a simple but effective method of sequence comparison [19]: Write two sequences along the two dimensions of a rectangle and place a dot wherever they are identical. When comparing a sequence to itself, this yields the main diagonal shown in Fig. 2.5. In our example, there are also two off-diagonals due to the repetition of TAT. In this section, we use dot plots to investigate the alcohol dehydrogenase locus in two species of the fruit fly, *Drosophila*.

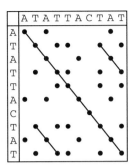

Fig. 2.5 Dot plot with matches of length three or more shown as lines

New Concepts

Name	Comment
dot plot	simple sequence comparison
gene duplication	evolutionary mechanism
orthology	result of speciation
paralogy	result of gene duplication

New Programs

Name	Source	Help
cchar	book website	cchar -h
dotPlotFilter.awk	book website	dotPlotFilter.awk -v h=1
randomizeSeq	book website	randomizeSeq -h
repeater	book website	repeater -h

Problem 127 Draw by hand a dot plot comparing $S = $ ACGTACGT to itself. Connect the dots along diagonals by lines. Can you explain the pattern?

Problem 128 Create a new working directory, DotPlot, and copy dmAdhAdhdup.fasta and dgAdhAdhdup.fasta into it. Which genes do these sequences encode, and which organisms are they taken from? Also, use cchar to count the nucleotides in each sequence.

Problem 129 In the following problems, we construct a pipeline for drawing a dot plot to compare dmAdhAdhdup.fasta and dgAdhAdhdup.fasta. How many cells will the dot plot contain?

Problem 130 Since dot plots display repeats between two sequences, we use our program repeater for finding repeats:

```
cat *.fasta | repeater
```

By default repeater returns the longest repeat. What is the longest repeat between dmAdhAdhdup.fasta and dgAdhAdhdup.fasta?

Problem 131 Use our program randomizeSeq to randomize the two example sequences. How long is the longest repeat now? Repeat the randomization a few times. Is it likely that the true longest repeat has occurred by chance?

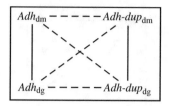

Fig. 2.6 All genes are characterized by homology (box); the solid lines connect pairs further characterized by orthology, the dashed lines pairs characterized by paralogy pairs by dashed lines

Problem 132 We now turn our attention to all repeats of some minimum length. When plotting these, we need to know the order in which the two sequences reach `repeater`. Hence we replace the `*.fasta` in our pipeline by

```
cat dmAdhAdhdup.fasta dgAdhAdhdup.fasta  |
repeater -m 12                           |
head -n 2
```

to get one example repeat of minimum length 12:

```
#len|strId:pos_1|...|strId:pos_n|seq
13|f2:3096|f1:3129|AAAATAGATAAAT
```

This output means that a repeat of length 13 has been found at position 3096 in sequence #2 (*D. guanche*) and at position 3129 in sequence #1 (*D. melanogaster*); the sequence of the repeat is AAAATAGATAAAT. To convert this output of `repeater` to dot plot coordinates, pipe it through the AWK program `dotPlotFilter.awk`. The usage of this program is explained in the header of the script.

Problem 133 Plot the output from `dotPlotFilter.awk` with various repeat lengths, upto 12; what do you observe?

Problem 134 We denote the four alcohol dehydrogenases we are dealing with by Adh_{dm}, $Adh\text{-}dup_{dm}$, Adh_{dg}, and $Adh\text{-}dup_{dg}$, where dm and dg refers to *D. melanogaster* and *D. guanche*, respectively. All four of them have a primeval *Adh*-sequence as their common ancestor, so they are covered by *homology* and often referred to as homologous. The genes that have diverged as a result of the duplication event are characterized by *paralogy* and often called paralogous. Genes that differ as a result of the divergence between *D. melanogaster* and *D. guanche* are characterized by *orthology* and usually called orthologous. Figure 2.6 depicts these classes of evolutionary relationships. Are the paralogous pairs or the orthologous pairs more similar? In other words, which of the two types of pairs has a more recent last common ancestor?

Problem 135 Draw by hand a cartoon phylogeny of the four *Adh* genes Adh_{dm}, $Adh\text{-}dup_{dm}$, Adh_{dg}, and $Adh\text{-}dup_{dg}$ (Fig. 2.6). Mark the times of duplication and speciation.

Problem 136 Look again at the dot plots from Problem 133. Can you spot the location of *Adh* and *Adh-dup*? Is the gene duplication visible in this plot?

Problem 137 One of the two species contains an insertion in the *Adh*-region. Can you spot it on the dot plot? Which species is affected?

Problem 138 To better understand our dot plot, we need to know the exon coordinates of *Adh* and *Adh-dup*. These are contained in the Genbank files `dmAdhAdhdup.gb` and `dgAdhAdhdup.gb`. Genbank files contain not only the sequence information but also annotations like exon positions. Look up the coordinates of the protein coding sequences (CDS) for the two genes *Adh* and *Adh-dup*.

Problem 139 Add the exons within the CDS of *D. melanogaster* as little boxes along the x-axis to our graph from Problem 133 (`repeater -m 12`). For example, an interval like 2021–2119 would become the coordinates (2012, 0), (2012, 150), (2119, 150), and (2119, 0). To achieve this, first write a pipeline to convert the CDS coordinates into a list of start and end positions, one pair per line, and save the output as `cdsDm.txt`. Then write `boxesX.awk` that takes as input `cdsDm.txt` and returns box-coordinates. Draw the box-coordinates together with the dot plot. Did the insertion affect an exon or an intron?

Problem 140 To further investigate the location of the insertion with respect to introns/exons, we draw lines across the graph to indicate the insertion. Lines are drawn in `gnuplot` as arrows without heads as follows:

```
set arrow from x1,y1 to x2,y2 nohead
```

Use this syntax to draw lines along the borders of the insertion in *D. melanogaster* across the graph. Since this adds more code to an already lengthy `gnuplot` command, save it in the script `cdsDm.gp` and use it

```
gnuplot -p cdsDm.gp
```

Problem 141 Write the script `boxesY.awk` for adding the exons of *D. guanche* along the y-axis, in addition to the exons of *D. melanogaster*. Then add horizontal and vertical lines along the CDS borders to see where they intersect with the lines of the dot plot. Summarize the `gnuplot` commands in the script `adhCds.gp`.

2.5 Optimal Alignment

Alignment algorithms are designed to reflect the homology relationship between the sequences analyzed. If the sequences are homologous across their entire lengths, a *global* alignment [37] is computed as shown Fig. 2.7a. If, on the other hand, they are homologous only locally, say just between coding exons, then they are analyzed using *local* alignment [44] as depicted in Fig. 2.7b. Notice that a global alignment between two sequences is simply one of very many possible local alignments between them. So global alignment is the generalization of local alignment. As a result, local

(a)   (b)

Fig. 2.7 Global (**a**) and local (**b**) homology between pairs of sequences. Homologous regions are shown in black and gray

alignment is used much more widely than global alignment. In this section, we move from dot plots to global alignment and then on to local alignment. The methods for global and local alignment are quite similar and are collectively referred to as "optimal alignment" because they always return the best result under a given score scheme.

New Concepts

Name	Comment
optimal global alignment	sequence comparison across full length
optimal local alignment	localized sequence comparison

New Programs

Name	Source	Help
cutSeq	book website	cutSeq -h
less	system	man less
time	system	man time

2.5.1   From Dot Plot to Alignment

Problem 142 Draw by hand a dot plot for sequences $S_1 = $ TTCAGGGTCC and $S_2 = $ TACAGTCC. Observe the convention that S_1 is written along the top horizontal edge and S_2 along the left vertical edge. Connect the dots in diagonally neighboring cells as follows:

Then write down a global alignment between S_1 and S_2 that maximizes the number of matched nucleotides. Which cell in the dot plot corresponds to the last column of the alignment?

Problem 143 How does the gap in the alignment appear in the dot plot?

2.5.2 Global Alignment

Problem 144 To compute global alignments, we extend the bottom up method for calculating the number of possible alignments and combine that with path-tracing in dot plots. Here is an alignment matrix for $S_1 = $ ACT and $S_2 = $ AC:

		-	A	C	T
		0	1	2	3
-	0				
A	1				
C	2				

An entry $F(i, j)$ in this matrix is located in row i and column j. Crucially, $F(i, j)$ is the score of the optimal alignment of the partial sequence $S_1[1..j]$ and $S_2[1..i]$. Which substrings of S_1 and S_2 does $F(2, 1)$ refer to? Name position and bases.

Problem 145 To actually calculate the values of $F(i, j)$, first define the score of two sequences of length zero as zero,

		-	A	C	T
		0	1	2	3
-	0	0			
A	1				
C	2				

Next, we fill in the top row, $F(0, j)$. To calculate $F(0, 1)$, we extend the "Null"-alignment in $F(0, 0)$ by the first nucleotide from S_1 and nothing, that is a gap, from S_2. Here we score a gap as -1. Hence we add -1 to the zero-score of the existing alignment and enter: The arrow points to the alignment we extended to calculate

		-	A	C	T
		0	1	2	3
-	0	0 $\leftarrow -1$			
A	1				
C	2				

this entry. Fill in the remainder of the first row of the alignment matrix.

Problem 146 Fill in the first column of the alignment matrix.

Problem 147 To determine $F(1, 1)$, go through the three possible extensions of an alignment and choose the best: Insertion of a gap into S_1 gives

$$\leftarrow -1 - 1 = -2$$

and similarly, insertion of a gap into S_2 gives

$$\uparrow -1 + -1 = -2.$$

The remaining possibility is extending the alignment by a nucleotide from S_1 and S_2. In our case this results in a match between two As. If we score match $= 1$, we can write

$$\nwarrow 0 + 1 = 1.$$

This is the best result so far, and hence our matrix entry. We can summarize these steps

$$F(i, j) = \max \begin{cases} F(i - 1, j) + g \\ F(i - 1, j - 1) + \text{score}(S_1[j], S_2[i]) \\ F(i, j - 1) + g \end{cases}$$

where g is the gap score; in our simplified algorithm we ignore gap opening, and hence the gap score is just the extension score, $g = g_e$. The boundary conditions for this recursion are used to fill in the first row and column of the alignment matrix as follows:

$$F(0, 0) = 0$$
$$F(0, j) = j \times g$$
$$F(i, 0) = i \times g$$

The only score still missing is the mismatch score, which we set to -1. Fill in the rest of the alignment matrix.

Problem 148 Once we have filled in the alignment matrix, the entry in the bottom right hand cell is the score of the best alignment possible given the score scheme. However, we do not yet know the actual alignment, only its score. To construct the corresponding alignment, follow the arrows from the lower right hand cell to the upper left hand cell. This is called "traceback" and creates the alignment from right to left as follows:

- $\nwarrow$: write the nucleotide from the horizontal sequence on top of the nucleotide from the vertical sequence;
- $\leftarrow$: write the nucleotide from the horizontal sequence on top of a gap;
- $\uparrow$: write a gap on top of the nucleotide from the vertical sequence.

Problem 149 Figure 2.8 shows the dynamic programming matrix for the global alignment of $S_1 = \text{TTCAGGGTCC}$ and $S_2 = \text{TACAGTCC}$ under the score scheme

- match $= 1$;
- mismatch $= -1$;
- gap, $g = -1$.

What is the score of the optimal alignment between S_1 and S_2?

$F(i,j)$		T	T	C	A	G	G	G	T	C	C
	0	1	2	3	4	5	6	7	8	9	10
0	0	←-1	←-2	←-3	←-4	←-5	←-6	←-7	←-8	←-9	←-10
T 1	↑-1	↖1	↖←0	←-1	←-2	←-3	←-4	←-5	↖←-6	←-7	←-8
A 2	↑-2	↑0	↖0	↖←-1	↖0	←-1	←-2	←-3	←-4	←-5	←-6
C 3	↑-3	↑-1	↖↑-1	↖1	←0	↖←-1	↖←-2	↖←-3	↖←-4	↖-3	↖←-4
A 4	↑-4	↑-2	↖↑-2	↑0	↖2	←1	←0	←-1	←-2	←-3	↖←-4
G 5	↑-5	↑-3	↖↑-3	↑-1	↑1	↖3	↖←2	↖←1	←0	←-1	←-2
T 6	↑-6	↖↑-4	↖-2	↑-2	↑0	↑2	↖2	↖←1	↖2	←1	←0
C 7	↑-7	↑-5	↑-3	↖-1	↑-1	↑1	↖↑1	↖1	↑1	↖3	↖←2
C 8	↑-8	↑-6	↑-4	↖↑-2	↖↑-2	↑0	↖↑0	↖↑0	↖↑0	↖↑2	↖4

Fig. 2.8 Dynamic programming matrix for aligning $S_1 = $ TTCAGGGTCC and $S_2 = $ TACAGTCC

Problem 150 When tracing back the path from the bottom right hand corner to the top left hand corner of the global alignment matrix in Fig. 2.8, there are cells with more than one arrow pointing back. In these cells the traceback path splits into two cooptimal alignments. Write down all cooptimal alignments implied by this matrix. What are their scores?

Problem 151 Make the directory OptimalAlignment. Change into it and construct two sequence files in FASTA format, seq1.fasta and seq2.fasta, containing $S_1 = $ TTCAGGGTCC and $S_2 = $ TACAGTCC. Then use the program gal to compute a global alignment of S_1 and S_2 under the score scheme defined in Problem 149. Does the alignment change if you vary the match and mismatch parameters?

Problem 152 Recall that gaps are scored as

$$g = g_o + l \times g_e,$$

where g_o denotes gap opening, l gap length, and g_e gap extension. Does the alignment change if you vary the gap opening and gap extension parameters?

Problem 153 In Problem 137 we used a dot plot to detect a large insertion in the *Adh-dup* of *D. melanogaster* when compared to *D. guanche*. Before we align *Adh-dup* from the two fruit flies to investigate this further, which of the two aligned sequences do you expect to contain a large gap?

Problem 154 Test the prediction made in Problem 153 by globally aligning dmAdhAdhdup.fasta and dgAdhAdhdup.fasta. View the result by piping it into the UNIX program less. You can navigate this using d for half a page down, and u for half a page up. Can you spot the region containing the large gap in *Adh-dup*? Press q to quit less. Look up the CDS coordinates in *.gb to determine which exon or intron is affected by the insertion.

2.5.3 Local Alignment

Problem 155 To get from global to local alignment, we change the algorithm, such that the score of an alignment cannot become negative. So, the first row and column are set to zero:

$$F(i, 0) = F(0, j) = 0$$

When filling in the remainder of the matrix, the maximum is formed as for the global alignment, but with zero included:

$$F(i, j) = \max \begin{cases} F(i - 1, j) + g \\ F(i - 1, j - 1) + \text{score}(S_1[j], S_2[i]) \\ F(i, j - 1) + g \\ 0 \end{cases}$$

Use these rules to compute by hand the local alignment matrix for $S_1 = \text{TACGT}$ and $S_2 = \text{GACGA}$ if $g = -1$, match $= 1$, and mismatch $= -1$.

Problem 156 The traceback for a local alignment starts at the greatest entry in the matrix and stops when the first zero is reached. Use this algorithm to determine the optimal local alignment of $S_1 = \text{TACGT}$ and $S_2 = \text{GACGA}$.

Problem 157 Use `lal` to compute the optimal local alignment between `dmAdhAdhdup.fasta` and `dgAdhAdhdup.fasta`. Compare the coordinates of the alignment to the CDS coordinates for *D. melanogaster* and *D. guanche*. What do you observe?

Problem 158 When comparing two sequences, there is only one global alignment, but there might be more than one local alignment (Fig. 2.7). Use `lal` to compute the best two local alignments between our two *Adh* files. This takes much longer than computing only the best local alignment. Use `time` to determine how much longer. Can you guess why computing two optimal local alignments is slow?

Problem 159 Which part of the *Adh/Adh-dup* region does the second best local alignment correspond to?

2.6 Applications of Optimal Alignment

Alignments are used whenever sequences are analyzed. Usually, the algorithms employed are faster versions of the optimal algorithms we have seen so far. However, in this section we survey two applications that demonstrate the potential usefulness even of slow optimal alignment. The first application is homology detection, the second dating the duplication of *Adh*.

	New Concepts
Name	Comment
homology detection	alignment compared to dot plot
divergence time	time since two genes split

2.6.1 Homology Detection

Problem 160 We have seen in Problem 136 that the homology between the paralogues *Adh* and *Adh-dup* is too weak to appear in the dot plot. Can we instead use global alignment to find a significant match between *Adh* and *Adh-dup* of *D. melanogaster*? We begin by looking up the coding sequences of both genes as follows:

```
grep CDS dmAdhAdhdup.gb
CDS join(2021..2119,2185..2589,2660..2926)
CDS join(3226..3321,3748..4152,4204..4521)
```

And then cut out the genomic region containing the CDS as follows:

```
cutSeq -r 2021-2926 dmAdhAdhdup.fasta > dmAdhCds.fasta
cutSeq -r 3226-4521 dmAdhAdhdup.fasta > dmAdhdupCds.
   fasta
```

Use gal to align dmAdhCds.fasta and dmAdhdupCds.fasta. What is the score of this alignment?

Problem 161 The score just found is rather low. Would a random alignment have a similar score? Use the program randomizeSeq to generate, say, ten scores for random alignments between the *Adh* and *Adh-dup*:

```
randomizeSeq -n 10 dmAdhCds.fasta |
gal -i dmAdhdupCds.fasta           |
grep Score
```

Use the program histogram to compute the distribution of 1000 random scores and plot it with gnuplot. Is the observed score likely due to chance?

Problem 162 We have seen that alignment finds traces of homology where dot plot does not. To make sure the alignment we found is reliable, let us try the converse: align two unrelated sequences and test the significance of their score. Save the first and last kb of dmAdhAdhdup.fasta into separate files, f1.fasta and f2.fasta. Align them and test the significance of the score by repeatedly randomizing, say, f2.fasta and realigning these randomized sequences with f1.fasta.

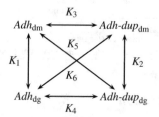

Fig. 2.9 The four genes in the *Adh/Adh-dup* region of *D. melanogaster* and *D. guanche*, and their pairwise substitution rates, *K*

2.6.2 Dating the Duplication of Adh

Problem 163 To date the *Adh* duplication, we calculate the six pairwise substitution rates between *Adh* and *Adhdup* from *D. melanogaster* and *D. guanche*. Since the speciation time for *D. melanogaster* and *D. guanche* is known to be approximately 32 million years [18], we can infer the duplication time by comparing the number of substitutions since speciation with the number of substitutions since duplication. The assumption we are making here is that the rate of substitution is constant throughout evolution, which is known as the "Molecular Clock" assumption. Figure 2.9 shows the four copies of *Adh* with the six substitution rates, $K_1, ..., K_6$ we wish to compute. Which of the substitution rates refer to speciation, which to gene duplication?

Problem 164 We base our estimates on the longest exon of *Adh* and *Adh-dup*, exon 2. Its coordinates are

Organism	*Adh*	*Adh-dup*
D. melanogaster	2185–2589	3748–4152
D. guanche	2145–2549	3540–3944

Use cutSeq to extract these sequences and call them dmAdhE2.fasta, dmAdhdupE2.fasta, and so on.

Problem 165 Align the *Adh* sequences from *D. melanogaster* and *D. guanche* and write a pipeline to compute the number of matches from this. Hint: The AWK function length(s) returns the length of string s.

Problem 166 The number of mismatches per site is called π. What is π between *D. melanogaster* and *D. guanche*?

Problem 167 The number of substitutions per site, or substitution rate, K, is a function of π [26]:

$$K = -\frac{3}{4} \log \left(1 - \frac{4}{3}\pi \right).$$

Compute the substitution rate between *Adh* from *D. melanogaster* and *D. guanche*, which is K_1 in Fig. 2.9.

Problem 168 What is the number of substitutions per site between *D. melanogaster* and *D. guanche* when estimated from exon 2 of *Adh-dup* (K_2 in Fig. 2.9)?

Problem 169 Having estimated the substitution rate for the split between *D. melanogaster* and *D. guanche*, we now estimate the substitution rate for the duplication, K_{dup}. For this purpose, compute the raw number of matches $M_3, ..., M_6$ implied by the remaining four sequence comparisons. Use their average to compute K_{dup}.

Problem 170 Given that the divergence time between *D. melanogaster* and *D. guanche* is known to be approximately 32 million years [18], how old is the duplication of *Adh*? Base your estimate on the average between K_1 and K_2. Draw a phylogeny with branch lengths proportional to the divergence times.

Chapter 3
Exact Matching

3.1 Keyword Trees

We often need to look up a short sequence, say $P =$ ACA, in a longer sequence, say $T =$ CACAGACACAT. This is known as the exact matching problem: to find all occurrences of a pattern P in a text T. For our example, the solution is illustrated here:

```
P    T
123 12345678901
ACA CACAGACACAT
```

P starts in T at positions 2, 6, and 8.

The simplest method for solving the exact matching problem is to align the start positions of P and T and to compare $P[1]$ and $T[1]$. If there is a mismatch, move P one position to the right and repeat, as shown in Fig. 3.1. In Step 1, a mismatch denoted by 0 is found when comparing $P[1]$ and $T[1]$. As a consequence, P is moved one step to the right and the matching resumed. This time all of P can be matched, and so on, right through to step 9, where $P[1]$ is mismatched with $T[10]$ and the algorithm stops, because P cannot be shifted further to the right. This is called the naïve string matching algorithm.

The speed of string matching is proportional to the number of comparisons. By counting all the zeros and ones in Fig. 3.1, we find that 16 comparisons were carried out. However, consider $P =$ AAA and $T =$ AAAAAAAAAAAA. Now P matches at every position in T leading to $10 \times 3 = 30$ comparisons. In fact, this is the worst case with run time, which is proportional to the product of the pattern and text lengths. Such proportionality is expressed using the big-O notation, which you can think of as "order of": $O(|P| \times |T|)$. This multiplicative run time can be avoided by realizing that it is not necessary to restart matching at the first position of P every time: After AAA has been found, the first two As have been matched already. In other words, the suffix $P[2..3]$ matches the prefix $P[1..2]$. So at the next step of the algorithm, only $P[3]$ should be compared to the current text position.

© Springer International Publishing AG 2017
B. Haubold and A. Börsch-Haubold, *Bioinformatics for Evolutionary Biologists*, https://doi.org/10.1007/978-3-319-67395-0_3

Step	1	2	3
T	CACAGACACAT	CACAGACACAT	CACAGACACAT
P	ACA	ACA	ACA
Match	0	111	0
Step	4	5	6
T	CACAGACACAT	CACAGACACAT	CACAGACACAT
P	ACA	ACA	ACA
Match	10	0	111
Step	7	8	9
T	CACAGACACAT	CACAGACACAT	CACAGACACAT
P	ACA	ACA	ACA
Match	0	111	0

Fig. 3.1 Naïve matching algorithm

Instead of searching for a single pattern, we might be looking for a whole set. Say we are looking for $P_1 = $ AAA and $P_2 = $ AAT; whenever AA is found, P_1 or P_2 could be detected in the next step. So we have made two observations for improving the naïve approach: suffixes in P might match prefixes of P, and prefixes of multiple patterns might match and hence ought to be summarized. Both of these observations lead to the preprocessing of one or more patterns into a tree, a keyword tree [3]. This guarantees an additive run time $O(|P| + |T|)$ and makes set matching fast.

In this section, we begin by implementing the naïve string matching algorithm in AWK. Then, we learn how to make it go faster.

New Concepts

Name	Comment
keyword tree	fast set matching
naïve string matching	simple but fast in many situations
set matching	matching multiple patterns

New Programs

Name	Source	Help
fold	system	man fold
gv	package manager	man gv
keywordMatcher	book website	keywordMatcher -h
latex	package manager	man latex
naiveMatcher	book website	naiveMatcher -h
revComp	book website	revComp -h
/usr/bin/time	system	man time

Problem 171 Begin as usual by creating a working directory, KeywordTrees, and changing into it. To solve the string matching problem in AWK, we need to address a string at specific positions. The AWK function split(s, a, fs) splits string s into array a on field separator fs and returns the number of fields. Use it to write the

AWK program `split.awk` that takes an arbitrary, short text from the command line using the syntax

```
awk -v t=CACAGACACAT -f split.awk
```

to assign a string to the variable `t`. The program then splits this string and iterates over the resulting array to print each character.

Problem 172 Write the program `naive.awk`, which takes a pattern P and a text T from the command line and prints out the starting positions of P in T. Use the AWK command `break` for jumping out of a loop, e.g.,

```
for(i=1; i<=n; i++)
   if (ta[i] == "X")
      break
```

Test your program on $T = \text{CACAGACACAT}$ and $P = \text{ACA}$.

Problem 173 To apply our program `naive.awk` to real sequences, we need to read sequences from FASTA files. AWK can execute system commands like `tail` and access the result by piping it through the AWK function `getline`:

```
BEGIN{
    cmd = "tail -n +2 " file
    while(cmd | getline)
        t = t $1
    print t
}
```

The function `getline` sets the AWK variables $1, 2, and so on as if the program were reading from a file. This could be run as

```
awk -f readFasta.awk -v file=mgGenome.fasta
```

Write `naive2.awk`, which reads a sequence from file and a pattern in FASTA format from stdin, and prints out the pattern's start positions. Where does ACGTCG occur in the genome of *M. genitalium*?

Problem 174 The program `revComp` computes the reverse complement of a sequence. Compute the reverse complement of `mgGenome.fasta`. Does it contain CGGCCT?

Problem 175 As we already said at the beginning, the naïve algorithm becomes slow when confronted with a pattern that matches everywhere, our example was $P = \text{AAA}$, $T = \text{AAAAAAAAAA}$. In that case, the run time is expected to be proportional to the product of the lengths of P and T, $O(|P| \times |T|)$. To explore the behavior of `naive2.awk` in this worst case, write a program that takes as input a sequence length, n, from the command line and writes a FASTA header followed by a single line of n As. Call the program `monoNuc.awk`. Run its output through the UNIX program `fold` to wrap the As into lines of length 80, which is the default output of `fold`.

Problem 176 Use monoNuc.awk and the UNIX-program time to get a run time for naive2.awk when applied to a text of 1 Mb and a pattern of 10 bp. What happens if you double the pattern and the text length? You can avoid printing all match positions to the screen by piping the output through tail.

Problem 177 Scripting languages like AWK tend to be slower than compiled languages like C. The program naiveMatcher is written in C and implements the same algorithm as naive2.awk. What is the run time of naiveMatcher compared to naive2.awk when searching for a 20-nucleotide pattern consisting of As in 2mb.fasta?

Problem 178 Draw the run time of naiveMatcher as a function of the length of the pattern consisting entirely of A when searching 2mb.fasta. Use pattern lengths of 10, 20, 50, 100, 200, 500, 1000, 2000, 5000, and 10000. Hint: Use

```
/usr/bin/time -p <command> 2>&1 | grep real
```

instead of plain time, to generate a table of run times ready for plotting. The command redirects the output of time from the error stream (2) to the standard output stream (1) and hence into the pipeline.

Problem 179 The naïve matching algorithm outlined in Fig. 3.1 can also be illustrated in a different way. Figure 3.2a shows the pattern to be matched as a graph consisting of nodes and edges depicted as arrows. Each node represents a state in the matching procedure and each arrow a response to match or mismatch. Match is illustrated by gray arrows, mismatch by orange arrows. The defining characteristic of the naïve algorithm is that upon every failure, matching resumes at the beginning of P; hence, all orange arrows in Fig. 3.2a point to the first node. However, a match to $P =$ ACA implies a match to the first character of P. Therefore, instead of returning to the beginning of P, the algorithm only needs to compare the characters from $P[2] =$ C onward. This is illustrated in Fig. 3.2b. What are the failure links for $P =$ AAA?

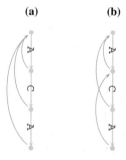

(a)   **(b)**

Fig. 3.2 A pattern to be matched shown as a graph. Gray arrows indicate matches, orange arrows "failure links" that are followed upon mismatch. Naïve failure links (**a**) always return to the beginning of the pattern. Better failure links (**b**) incorporate the fact that after the last A has been matched, the first A has also been matched already

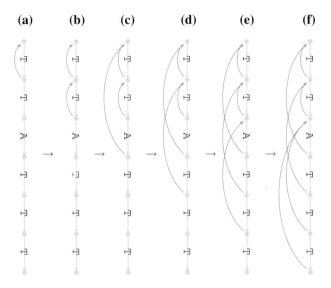

Fig. 3.3 Systematic construction of failure links going from the initialization (**a**) through to the fully preprocessed pattern (**f**)

Problem 180 To systematically construct failure links, we begin by stating that a mismatch after the first match means: return to the beginning (Fig. 3.3a). After this initialization step, we work our way from top to bottom by following failure links until the earliest match to the character directly above the node to be connected. The node we reach is the target of the next failure link. For example, in Fig. 3.3b we look for a match to T; after following the existing failure link, we follow the match link and have thus found the target for the failure link. Next, two existing failure links are followed without a match, and hence the new failure link points to the start node (Fig. 3.3c). Then, another failure followed by a match in Fig. 3.3d, and this pattern is repeated in Fig. 3.3e. Finally, two failure links need to be followed before a match is found leading to the completely preprocessed pattern in Fig. 3.3f. Construct the failure links for $P = $ ATATAT.

Problem 181 The program keywordMatcher looks for exact matches after preprocessing the pattern with failure links. Its name comes from the fact that the preprocessed pattern in Fig. 3.2 is called a "keyword tree". What is the run time of keywordMatcher when searching for a string of 20 A in 2mb.fasta? Compare your result to the corresponding run time of naiveMatcher.

Problem 182 Apply keywordMatcher to the same increasingly long patterns used in Problem 178 and plot the results in the same graph as the times for naiveMatcher.

Problem 183 Compare the run times of naiveMatcher and keywordMatcher when searching a random sequence of length 20, say

$$P = \text{TTTAACCTCCGGCGGAGTTT}$$

in random sequences of lengths 1, 2, 5, 10, 20, 50, 100 Mb (ranseq). Plot the results
in a single graph.

Problem 184 Keyword trees were originally developed for set matching [3] and
the program keywordMatcher can search for many patterns simultaneously. Say,
we wish to look for five patterns (keywords): $P_1 = $ ACG, $P_2 = $ AC, $P_3 = $ ACT,
$P_4 = $ CGA, and $P_5 = $ C. Notice that P_2 is contained in P_1 and P_3, P_5 in all others,
and P_1 and P_3 have the matching prefix AC. To make searching efficient, matching
prefixes are summarized as common paths in the keyword tree. Its construction is
shown in Fig. 3.4. The patterns are sequentially fitted into a growing tree structure.
The first partial tree for $P_1 = $ ACG in Fig. 3.4b should look familiar from our graphs
for single patterns, except for the label on the end node indicating the pattern just
matched.

Repeat the keyword tree construction using paper and pencil. Then, initialize the
failure link construction as in Fig. 3.4f and enter the five missing failure links.

Problem 185 Keyword trees can be drawn automatically using the -t option of
keywordMatcher, for example,

```
keywordMatcher -t kt.tex -p ACA mgGenome.fasta > /dev/
    null
```

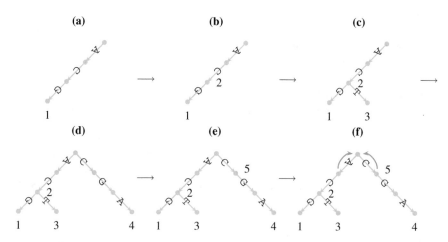

Fig. 3.4 Sequential construction of a keyword tree for the five patterns $P_1 = $ ACG, $P_2 = $ AC,
$P_3 = $ ACT, $P_4 = $ CGA, and $P_5 = $ C

writes the tree for ACA to the file kt.tex and discards the output by writing it to the null device. The file kt.tex contains code in the typesetting language LATEX [28, 30]. To view this, apply the program latex to the wrapper file for kt.tex, ktWrapper.tex. Like kt.tex, it is generated automatically and contains LATEX commands. If you have never used LATEX, take a look at ktWrapper.tex. Commands start with a backslash character. For example, the line

```
\input{kt.tex}
```

imports the keyword tree contained in the file kt.tex. Apart from commands, you can also enter ordinary text in a LATEX document (try this). Like many programming languages, LATEX needs to be compiled to generate the desired output, a neatly typeset page:

```
latex ktWrapper.tex
```

This generates the file ktWrapper.dvi, which is then converted to the postscript file ktWrapper.ps

```
dvips ktWrapper.dvi
```

ready for viewing:

```
gv ktWrapper.ps &
```

Automatically generate the keyword tree for our five example patterns $P_1 = $ ACG, $P_2 = $ AC, $P_3 = $ ACT, $P_4 = $ CGA, and $P_5 = $ C. Also, what happens if you delete the line

```
\date{}
```

in ktWrapper.tex and run LATEX again?

Problem 186 Use the keyword tree just constructed to trace with pencil and paper the search for P_1–P_5 in $T = $ ACGC. The rule is, whenever a labeled node is reached, the corresponding pattern has been found. However, that rule is not sufficient for finding all patterns. Can you see where it fails, and how this might be fixed?

Problem 187 Write down the output set next to each node on our example keyword tree. Each output set contains at least the current label but perhaps also additional elements.

3.2 Suffix Trees

Think of your favorite book, say the first volume of *Harry Potter*. How would you find every passage that contains the word "Voldemort"? You could scan every page. But there are many repeated words and a few repeated phrases that you would scan again and again. So, would not it be useful if you could compress the book such that every repeat is only listed once together with its occurrences? An index is such a compressed version of a book. Novels usually do not have one but textbooks usually do. However, even the most detailed index does not contain every word, because readers do not need to look up insignificant words like "and". Still, sometimes it is desirable to have an index of every possible word.

A suffix tree of a text is an index that references not only every word but every possible substring [21]. Consider as text ADAM. To construct the corresponding suffix tree, start by drawing a root and a leaf node connected by an edge (Fig. 3.5a). Label the edge with the first suffix, ADAM, and the label with 1 to indicate the end of the first suffix (Fig. 3.5b). Next, take the suffix starting at position 2, DAM, and fit it into the tree starting from the root. Since its first character, D, already mismatches the first A of ADAM, create a new branch and label it as before (Fig. 3.5c). The third suffix, AM, can be fitted one step into the tree before a mismatch occurs; apart from a leaf, this also creates an internal node (Fig. 3.5d). The last suffix, M, again branches directly off the root (Fig. 3.5e).

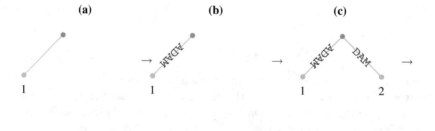

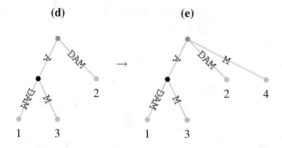

Fig. 3.5 Construction of the suffix tree that indexes ADAM. The root is shown in brown, leaves in green, and the single internal node in black

If you now look for the occurrence of A in ADAM, start searching at the root of the suffix tree. The leaf labels below the match, in this case 1 and 3, indicate the positions of A in ADAM. This means that—like a book index—a suffix tree enables us to look up a word in time proportional to the length of the pattern. But unlike a book index, a suffix tree references *all* possible words, or substrings, of a text. Surprisingly, a suffix tree not only provides a perfect index but it can also be computed efficiently, as we explain in this section.

New Concepts

Name	Comment
longest repeat	found using suffix tree
shortest unique substring	becomes repeat when trimmed on the right
suffix tree	text index

New Program

Name	Source	Help
shustring	book website	shustring -h

Problem 188 As we saw in Fig. 3.5, a suffix starts somewhere in a sequence and ends at its end. Consider the sequence $S = $ ACCCG and write down its suffixes.

Problem 189 Write down the suffix tree for S by following the procedure illustrated in Fig. 3.5.

Problem 190 Use the suffix tree constructed in Problem 189 to search for CC in S by matching from the root as explained for Fig. 3.5.

Problem 191 Suppose our sequence mutates at its last position from a G to an A, such that now $S = $ ACCCA. Write down the suffix tree for this sequence. What do you observe?

Problem 192 Write down the suffix tree for $S = $ ACCCA$.

Problem 193 The concatenated strings along the path leading from the root of the suffix tree to a node is called the node's "path label". The length of this path label is the node's "string depth". Add string depths to the internal nodes of the suffix tree constructed in Problem 189.

Problem 194 Every path label that leads to an internal node is a repeated string. Use the suffix tree with string depths constructed in Problem 193 to find the longest repeat in S.

Problem 195 Make the directory SuffixTrees, change into it, and copy the genome of *M. genitalium* contained in mgGenome.fasta from Data to your current directory. The program repeater computes a suffix tree of the input sequence to find longest repeats. How long is the longest repeat in the genome of *M. genitalium*?

Problem 196 Next, we compute the length of the longest repeat expected in a random version of the *M. genitalium* genome. We begin by calculating the probability of a match between two random nucleotides. Since there are four possibilities of obtaining a match, AA, CC, TT, and GG, the probability of randomly drawing two identical nucleotides from a sequence in which each nucleotide has frequency 1/4 is $1/16 \times 4 = 1/4$. However, in real sequences, the nucleotides usually do not occur with equal frequencies. Use the program cchar to compute the nucleotide composition of *M. genitalium* (mgGenome.fasta). What is the probability of drawing AA when picking two random nucleotides from the genome of *M. genitalium*?

Problem 197 Write a pipeline to calculate the probability of drawing two identical nucleotides from the genome of *M. genitalium*.

Problem 198 To get from the match probability, P_m, to the expected length of the longest repeat in the genome of *M. genitalium*, consider a toy dot plot with three matches, two of length 1 and one of length 2:

The probability of drawing a dot is P_m. The probability of drawing a diagonal of length l, and hence a match of length l, is P_m^l. The expected number of such diagonals is their probability times the number of cells in the dot plot. When comparing two sequences of length L, this is $(L - l)^2$, which is approximately L^2. Hence, the expected number of l-mer matches, n_e, is

$$n_e = P_m^l \times L^2.$$

Since we are looking for the *longest* such match, we set $n_e = 1$. What is the expected length of the longest repeat in the genome of *M. genitalium*? How does this compare to the observed longest repeat?

Problem 199 To check the expected match length computed in Problem 198, randomize the sequence of *M. genitalium* using the program randomizeSeq and compute the longest repeat in that randomized version. Do this a few times and compare your results to the expectation.

Problem 200 Most books come without an index. One reason for this is that making an index is a lot of work. We have already seen that repeater is quick when applied to the genome of *M. genitalium*. However, with just half a megabase, this genome is very small compared to, say, ours with 3.2 gigabases. So the question is, how time-consuming is suffix tree construction in general? Consider the "naïve" construction method depicted in Fig. 3.5. How does its run time scale as a function of sequence length? Approach this question by first constructing a suffix tree for a sequence consisting of only one type of nucleotide, for example, AAAA.

Problem 201 The program repeater uses a more efficient algorithm than the naïve construction in Fig. 3.5. To investigate how the more sophisticated algorithm scales with sequence length, use the program ranseq to simulate sequences of 1, 2, 5, 10, and 20 Mb. Apply repeater to them, measure its run time with time, and plot it. What is the run time of repeater?

Problem 202 Repeat the resource analysis of repeater, but instead of time, measure memory using commands like

```
ranseq -l 1000000 | /usr/bin/time -f "%M kb" repeater
```

on Linux or

```
ranseq -l 1000000 | /usr/bin/time -l repeater
```

on Mac OS X.

3.3 Suffix Arrrays

The nodes of suffix trees consume a lot of computer memory. This becomes a problem when computing suffix trees for very long sequences such as mammalian genomes with their billions of nucleotides. Hence, a space-saving alternative to suffix trees has been developed, which consists solely of the sorted array of suffixes [36].

Problem 203 To compute a suffix array, we again begin by writing down every suffix of the input sequence. But this time we automate the procedure. Save our sequence $S = $ ACCCA$ in the file s.fasta. Then, copy the program

```
!/^>/{
    s = s $1; # read the input sequence
}END{
    L = length(s);
    for(i=1; i<=L; i++)
        print i "\t" substr(s, i);
}
```

into the file suf.awk and use it to generate the suffixes of S and their starting positions. Next, sort the suffixes alphabetically. Finally, number the lines in your output. Compare your result to the suffix array shown in Fig. 3.6a.

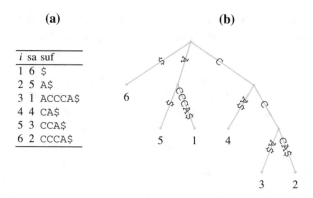

Fig. 3.6 Suffix array (**a**) and corresponding suffix tree (**b**) of ACCCA$

New Concepts

Name	Comment
enhanced suffix array	suffix array plus lcp array
inverse suffix array	suffixes in textual order
lcp array	lengths of longest common prefixes in a suffix array
lcp interval tree	suffix tree from enhanced suffix array
suffix array	sorted list of suffixes

Problem 204 Figure 3.6b shows the suffix tree for the sequence ACCCA$, which should look familiar from Chap. 3.2. It is closely related to the suffix array next to it. Take the internal node with path label A. The leaves connected to that node refer to suffixes 5 and 1. These are neighbors in the suffix array, where they occupy entries sa[2] = 5 and sa[3] = 1. We denote this interval sa[2..3]. Write down the suffix array intervals that correspond to the two remaining internal nodes and the root.

Problem 205 To make the close relationship between suffix array and suffix tree more explicit, we add to our suffix array the array cp for "common prefix", where cp[i] is the longest common prefix between sa[i] and sa[$i-1$]. For example, suf[6] in Fig. 3.6a is CCCA$, and suf[5] = CCA$; since these two suffixes have the common prefix CC, cp[6] = CC. Add cp to the suffix array and mark the prefixes in the suffix tree. The first suffix in sa cannot be compared to another suffix; hence, it is "not defined", nd.

Problem 206 To reconstruct the suffix tree from its suffix array, we need the lengths of the common prefixes, rather than the prefixes themselves. Add the lcp array to the suffix array, where lcp[i] is the length of cp[i].

Problem 207 The distinct entries in the lcp array computed in Problem 206, 0, 1, and 2, correspond to the string depths in the suffix tree in Fig. 3.6. Our aim is still to reconstruct that tree by traversing the suffix array enhanced by the lcp array. For this purpose, we first extend the lcp array by a "stop" entry, -1, as shown in Fig. 3.7a. In addition, we write next to the lcp array an empty table with the three distinct string

(a)

index	sa	lcp	2 1 0
1	6	-1	· · ·
2	5	0	· · ·
3	1	1	· · ·
4	4	0	· · ·
5	3	1	· · ·
6	2	2	· · ·
7	–	-1	· · ·

(b)

index	sa	lcp	2 1 0
1	6	-1	────
2	5	0	· · ·
3	1	1	· · ·
4	4	0	· · ·
5	3	1	· · ·
6	2	2	· · ·
7	–	-1	· · ·

(c)

index	sa	lcp	2 1 0
1	6	-1	────
2	5	0	────
3	1	1	· · ·
4	4	0	· · ·
5	3	1	· · ·
6	2	2	· · ·
7	–	-1	· · ·

(d)

index	sa	lcp	2 1 0
1	6	-1	────
2	5	0	──┐
3	1	1	──┘
4	4	0	· · ·
5	3	1	· · ·
6	2	2	· · ·
7	–	-1	· · ·

Fig. 3.7 The enhanced suffix array with an auxiliary table for reconstructing the corresponding suffix tree

depths 2, 1, and 0 as headers. Now we traverse the lcp array to find the intervals in the sa array that corresponds to the suffix tree. Starting from the top, check at each position the relationship between the current entry in the lcp-array, lcp[i] and the next entry, lcp[i + 1]:

- If lcp[i] < lcp[i + 1], open one or more intervals. The number of intervals to open is lcp[i + 1] − lcp[i]. The string depths of the opened intervals are the values between lcp[i] + 1 and lcp[i + 1]. In our example, lcp[1] = −1 and lcp[2] = 0; since lcp[1] < lcp[2], we open 0 − −1 = 1 interval with string depth −1 + 1 = 0. To denote this, we draw the gray line in Fig. 3.7b. Similarly, in the next step, we observe lcp[2] < lcp[3] and open another lcp interval, as shown by the red line in Fig. 3.7c.
- If lcp[i] > lcp[i + 1], close one or more intervals where the string depth is greater than lcp[i + 1] and has occurred in the interval to be closed. In our example, we observe in Fig. 3.7d that lcp[3] > lcp[4]. Since the string depth of the red, but not of the gray interval, is greater than lcp[4], and it occurs in the interval under consideration, the red interval gets closed, while the gray interval remains open.

Find the remaining lcp intervals.

Problem 208 Convert the nested structure of the lcp intervals in Problem 207 to a tree, the lcp interval tree [38, p. 85ff]. Each node has the format $\ell - [i..j]$, where ℓ is the string depth, i is the start index, and j is the end index.

Problem 209 Let us study the relationship between enhanced suffix array and suffix tree with a longer example sequence, $S = \text{GTAAACTATT}$. Write down its enhanced suffix array, the nested lcp intervals, and the suffix tree.

Problem 210 As we have seen, lcp arrays allow the conversion of suffix arrays to suffix trees. The efficient computation of lcp arrays is therefore central to the application of suffix trees. In the following couple of problems, we learn how to do this. The crucial insight is that the lcp value for a suffix $S[i..]$ forms a lower bound of the lcp value of the suffix one step to the right, $S[i + 1..]$. Consider, for example, the sequence ACCCACG. Its first suffix matches AC at position 5; from this, we can conclude that its second suffix matches at least the C at position 6 from the previous match. In other words, if ℓ is the length of the common prefix of $S[i..]$, then $\ell - 1$ is the lower bound of the lcp for $S[i + 1..]$. The emphasis here is on *lower bound*; in our example, the CC at the beginning of the second suffix match is the CC at the beginning of the third. To use this lower bound insight, we need to traverse the suffix array in the same order in which the suffixes appear in the input sequence. The mapping between sa and S is called the inverse suffix array, isa, which is defined as

$$\text{isa}[\text{sa}[i]] = i.$$

Add the isa to the suffix array of ACCCA\$. This is most effective when done to the left of the index i.

Problem 211 Write the program isa.awk that takes as input the sorted output from suf.awk and adds the isa as another column.

Problem 212 Algorithm 2 shows how to compute the lcp values in linear time [27]. The inverse suffix array, isa, is computed in lines 1–2, which should look familiar from Problem 211. Implement this algorithm in a program esa.awk that takes as input the sorted output of suf.awk and prints the enhanced suffix array, that is, sa and lcp. Hint: The AWK function

```
substr(s, p, n)
```

returns the substring of s that starts at p and is n characters long.

Problem 213 Now, we apply our program to real sequences. What is the longest repeat sequence in *Adh/Adh-dup* of *Drosophila guanche* (dgAdhAdhdup.fasta)?

Problem 214 What is the longest repeat sequence in *Adh/Adh-dup* of *Drosophila melanogaster* (dmAdhAdhdup.fasta)?

Problem 215 By typing

```
cat dmAdhAdhdup.fasta dgAdhAdhdup.fasta > dmDgAdhAdhdup
    .fasta
```

the two alcohol dehydrogenases we just investigated can be written into the same file. What is the longest repeat in the concatenated sequences? Its length will be greater than that of the repeats within the individual sequences. Can you think of why?

Algorithm 2 Algorithm for computing the lengths of common prefixes [27]

Require: S {input sequence}
Require: n {length of S}
Require: sa {suffix array}
Ensure: lcp {array of lengths of longest common prefixes}
1: **for** $i \leftarrow 1$ to n **do**
2:    isa[sa[i]] $\leftarrow i$ {construct inverse sa}
3: **end for**
4: lcp[1] $\leftarrow -1$ {initialize lcp}
5: $\ell \leftarrow 0$
6: **for** $i \leftarrow 1$ to n **do**
7:    $j \leftarrow$ isa[i]
8:    **if** $j > 1$ **then**
9:        $k \leftarrow$ sa[$j-1$]{$S[k..]$ is left-neighbor of $S[i..]$ in sa}
10:       **while** $S[k+\ell] = S[i+\ell]$ **do**
11:           $\ell \leftarrow \ell + 1$
12:       **end while**
13:       lcp[j] $\leftarrow \ell$
14:       $\ell \leftarrow \max(\ell - 1, 0)$ {ℓ cannot become negative}
15:   **end if**
16: **end for**

Problem 216 In a suffix tree, repeats are path labels ending above internal nodes. If instead we look at the path labels of leaves, we find unique sequences. For example, CCC in Fig. 3.6 is unique, as it is found on the edge connected to leaf 2. The extension of this motif to CCCA does not change its property of uniqueness as we are still on the path to leaf 2. Hence, when talking about unique motifs, we are particularly interested in *shortest* motifs, which we call shortest unique substrings, or shustrings. By traversing all leaves, the shustrings starting at every position in a sequence can be determined: At each leaf, extend its parent's path label by a single nucleotide. Write down the shustrings for each position in ACCCA. What is the length of the shortest shustring(s)? What might be the use of finding such shortest shustrings in real sequences?

Problem 217 Instead of using the suffix tree of ACCCA, we can also use the lcp array to look up the shustring lengths at every position. How is this done?

Problem 218 The program `shustring` implements the search for shustrings based on suffix arrays. What is the length of the shortest unique substrings in the genome of *M. genitalium*?

Problem 219 By default, `shustring` only searches the forward strand. Does the result change when the reverse strand is included?

Problem 220 Instead of looking for the globally shortest shustrings, you can also look for shustrings of a certain minimal (-m) or maximal (-M) length. This requires the *local* option, which is invoked for a particular sequence; in our case, there

is only one sequence to choose from, so anything that matches the header line in
mgGenome.fasta will do the trick, for example

-1 .

because in a regular expression a dot matches anything. How many shustrings of
length ≤ 7 are contained in the genome of *M. genitalium*?

3.4   Text Compression

Compressed	Starting Sequence	Sorted	Compressed
$A_4C_2T_5C_2$ $\leftrightarrow$ AAAACCTTTTTCC	$\rightarrow$ AAAACCCCTTTTT $\leftrightarrow$	$A_4C_4T_5$

Fig. 3.8 Compressing sequences

The ability to compress data underlies many powerful programs ranging from mp3
players to read mappers in Bioinformatics. Given the starting sequence in Fig. 3.8, it
can be compressed by counting repeated nucleotides and writing them as, for example, A_4. This kind of compression is easy to reverse, though only effective if there are
long runs of identical nucleotides. Sorting the sequence would maximize such runs,
which would lead to the greatest compression. However, sorting is irreversible. The
Burrows–Wheeler transform was devised to reversibly increase the runs of identical
symbols in any type of data [12]. We begin by transforming the word "Mississippi":

```
              111
      123456789012
      mississippi$
```

It consists of eleven characters plus a sentinel at the end. The first step in the transform
is to write down all the rotations of the word

```
 1 mississippi$
 2 ississippi$m
 3 ssissippi$mi
 4 sissippi$mis
 5 issippi$miss
 6 ssippi$missi
 7 sippi$missis
 8 ippi$mississ
 9 ppi$mississi
10 pi$mississip
11 i$mississipp
12 $mississippi
```

Next, the rotations are sorted:

String	Rotations	Sorted Rotations	Transformed String
	mississippi$	$mississippi	
	ississippi$m	i$mississipp	
	ssissippi$mi	ippi$mississ	
	sissippi$mis	issippi$miss	
	issippi$miss	ississippi$m	
mississippi$	ssippi$missi	mississippi$	ipssm$pissii
	sippi$missis	pi$mississip	
	ippi$mississ	ppi$mississi	
	ppi$mississi	sippi$missis	
	pi$mississip	sissippi$mis	
	i$mississipp	ssippi$missi	
	$mississippi	ssissippi$mi	

The first column of the sorted rotations consists of the characters in Mississippi in alphabetical order. As we have already observed, this cannot be decoded to the original word. But intriguingly, the last column can, and this is our transform.

How is this transform reversed, that is, decoded? We start by sorting the transform to give us the first column of the sorted rotations, column $\mathscr{F}$ in Fig. 3.9a, while the transform is in column $\mathscr{L}$. Then we label each character in $\mathscr{F}$ and $\mathscr{L}$ with its count, so the first i is labeled i_1, the second i_2, etc. (Fig. 3.9b). Finally, we look for the sentinel in $\mathscr{L}$, jump across to $\mathscr{F}$, and have found the first character of the original string, m_1, in our example (Fig. 3.9c). Next, we look up m_1 in $\mathscr{L}$, and jump across to $\mathscr{F}$ to find our second character, i_4, and so on until this traceback yields the original "mississippi$". In this section we explore the effect of BWT and how it can be used to compress sequences.

New Concepts

Name	Comment
Burrows–Wheeler transform	Reversible sorting
Compressibility	Opposite of complexity
Lempel–Ziv decomposition	Measure sequence complexity
Move to front	Reduce sequence complexity

New Programs

Name	Source	Help
bwt	Book website	bwt -h
bzip2	System	man bzip2
du	System	man du
gzip	System	man gzip
lzd	Book website	lzd -h
mtf	Book website	mtf -h

Problem 221 Write down the Burrows–Wheeler transform of the sequence TACTA on a piece of paper. Create the directory Bwt and change into it. Check your manual transform using the program bwt.

(a)		(b)		(c)	
$\mathcal{F}$	$\mathcal{L}$	$\mathcal{F}$	$\mathcal{L}$	$\mathcal{F}$	$\mathcal{L}$
\$	i	$\$_1$	i_1	$\$_1$	i_1
i	p	i_1	p_1	i_1	p_1
i	s	i_2	s_1	i_2	s_1
i	s	i_3	s_2	i_3	s_2
i	m	i_4	m_1	i_4	m_1
m	\$	m_1	$\$_1$	m_1	$\$_1$
p	p	p_1	p_2	p_1	p_2
p	i	p_2	i_2	p_2	i_2
s	s	s_1	s_3	s_1	s_3
s	s	s_2	s_4	s_2	s_4
s	i	s_3	i_3	s_3	i_3
s	i	s_4	i_4	s_4	i_4

Fig. 3.9 Decoding the Burrows–Wheeler transform: write down the first ($\mathcal{F}$) and last ($\mathcal{L}$) columns of the rotations **a**; count characters **b**; trace back **c**, showing the first two steps

Problem 222 To observe the effect of BWT on natural language, we next transform Shakespeare's *Hamlet*, which is contained in `hamlet.fasta`. Begin by looking at the file to convince yourself that it contains the bard's words, albeit in a slightly unorthodox format:

```
less hamlet.fasta
```

When you press d, `less` moves down by half a page, u does the opposite, and q lets you quit. One advantage of `less` is that text can be piped into it. For example, the BWT of *Hamlet* is

```
bwt hamlet.fasta | cat -n | less
```

Take a look at the first couple of pages. What do you observe?

Problem 223 Decode by hand the BWT CCCT\$G. Check the result using `bwt`.

Problem 224 We have seen that decoding a BWT is relatively simple, but encoding is tedious due to the rotation step. However, by using a suffix array we can avoid the rotations. To see how this works, return to our original example, $T = $ `mississippi$` and write down its suffix array using the program `suf.awk` from Problem 203. Notice that up to the sentinel rotations and suffixes are identical. Can you think of a method for finding the BWT by traversing the sa?

Problem 225 Construct the suffix array of $T = $ TACTA\$ and use it to infer the BWT of T. Check your result using `bwt`.

3.4.1 Move to Front (MTF)

To make the BWT more compressible, we can apply the move to front (MTF) procedure [2]. This works by noting the position in the alphabet of a given character. So to
encode the nucleotide sequence $S = $ GTTT, we use as alphabet the four nucleotides:

$$0 \; 1 \; 2 \; 3$$
$$\text{A C G T}$$

Characters are encoded by their position in the alphabet and this position can change
after each encoding step: The first G in S is encoded as 2; in the next iteration, the G
is moved to the front of the alphabet

$$0 \; 1 \; 2 \; 3$$
$$\text{G A C T}$$

and the T is a 3. Now the T is moved to the front

$$0 \; 1 \; 2 \; 3$$
$$\text{T G A C}$$

The remaining two T are encoded as 0, yielding 2, 3, 0, 0. To decode this, we reverse
the procedure. Start with ACGT; the first 2 stands for G, rearrange alphabet to GACT;
the 3 stands for T, rearrange alphabet to TGAC; the two zeros stand for T, and we
have GTTT back. The crucial idea of MTF is that repeats of *any* nucleotide lead to
runs of zeros, which can then be compressed.

Problem 226 Use MTF to manually encode $T = $ GTTAG. Check the result using
mtf.

Problem 227 Let 1,0,1,3,3 be the result of MTF on the alphabet ACGT. Decode the
original string.

3.4.2 Measuring Compressibility: The Lempel–Ziv
Decomposition

In order to measure the effect of BWT and MTF, we need to quantify the extent to
which a transform can be compressed, compared to the original sequence. One way
to do this is to look for matching regions or "factors" within the input sequence. The
fewer such match factors can be found, the more compressible the sequence is. Take
for example the sequence $S = $ CCCG; at every position i in S we ask, how long is
the longest match that appears somewhere to the left of i? The first position has no
left neighbor, so the rule for no match is invoked: $S[i]$ becomes a factor, C, and we
move to the next position $S[i + 1]$. This matches the first position up to the forth,
so our second factor is CC. The last position, $S[4] = $ G, again has no match, so S is
decomposed into the three factors

C.CC.G

This decomposition is called the Lempel–Ziv decomposition [32] and is widely used in compression programs, for example in `gzip`.

Problem 228 Calculate by hand the Lempel–Ziv decomposition of TACTA. Check your result using the program `lzd`.

Problem 229 Calculate by hand the Lempel–Ziv decomposition of the maximally redundant sequence AAAAA. Again use `lzd` to check your result.

Problem 230 Instead of directly measuring compressibility, it is often simpler to measure its opposite, which is complexity. Highly complex sequences cannot be compressed much, and vice versa. We define as complexity the number of Lempel–Ziv factors divided by sequence length:

$$C = \frac{\text{number of LZ factors}}{|S|}.$$

What are the complexities of the example sequences in Problems 228 and 229?

Problem 231 What are the theoretical limits of C?

Problem 232 In practice, the largest value C can take is the number of LZ factors per nucleotide in a random sequence. Use `ranseq` to generate random sequences of lengths 1, 2, 5, and 10 kb, and measure C. Plot C as a function of sequence length. Is the maximum of C constant with respect to sequence length?

Problem 233 Next, we compute C for the genome of *M. genitalium* to observe the effect of the Burrows–Wheeler transform, BWT, and move to front, MTF, on complexity. Fill in the table below:

Operation	Number of Factors	C
None		
BWT		
BWT | MTF		
MTF		
MTF | BWT		

Which combination gives the largest reduction in C?

Problem 234 We return to compression, the opposite of complexity. Two examples of popular compression programs are `gzip` and `bzip2`. The "zip" in these program names refers to Lempel–Ziv, the "b" in `bzip2` to BWT. Measure the size of `mgGemone.fasta` by using the program `du` for disk usage:

```
du -h mgGenome.fasta
```

Then compute the size of the compressed genome after it has undergone randomization, BWT, and MTF. Summarize your results in the following table:

Operation	gzip bzip2
Nothing	
randomizeSeq	
BWT	
BWT | MTF	

Chapter 4
Fast Alignment

4.1 Alignment with k Errors

Exact matching is fast and uses little memory, while filling in alignment matrices is slow and memory consuming. Of these two central resources, time and memory, memory is often the more limiting: it is usually possible to wait a bit longer for a computation to finish, while a computation that does not fit into memory cannot even start. However, waiting for a calculation to complete is also undesirable, not least because computations that finish "instantaneously" can become part of larger, new applications. Useful programs are efficient.

The central insight for speeding up alignment is that the vast majority of cells in an alignment matrix are nowhere near an optimal path. But how can the "promising" parts of an alignment matrix be identified? The answer is, cheap exact matching. The k-error alignment method makes direct use of this idea [8]: Say we are looking for a short query sequence, Q, in a long subject sequence, S, and Q contains a single error. This could be an insertion, a deletion, or a mismatch; in the example shown in Fig. 4.1a it is a deletion, marked by a Δ.

If Q is divided into two fragments, a and b, the error is either located in segment a, as in Fig. 4.1a, or b, but not in both. Next, the algorithm searches for a and b in S and finds b (Fig. 4.1b). This match anchors the alignment matrix, which occupies $(|Q| + 1)^2$ cells rather than $(|Q| \times |S|)$ (Fig. 4.1c).

Put more generally, k-error alignment begins by choosing an upper limit for the number of errors the final alignment may contain, k. Then Q is divided into $k + 1$ fragments of equal length, each of which is searched exactly in S. Whenever a match is found, an alignment matrix is constructed with dimension $O((|Q| + k)^2)$, which is filled in by dynamic programming to check whether it contains a k-error match between Q and S.

As k grows, the fragment size decreases. Short fragments can match random positions in the subject making the checking phase potentially *more* time consuming than filling in the full alignment matrix. However, for small k the method works well in practice as we shall see in this section.

© Springer International Publishing AG 2017
B. Haubold and A. Börsch-Haubold, *Bioinformatics for Evolutionary Biologists*, https://doi.org/10.1007/978-3-319-67395-0_4

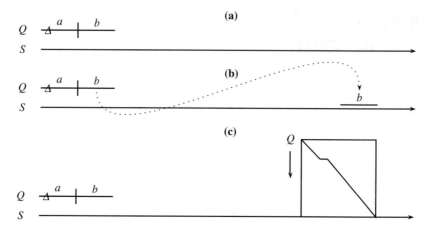

Fig. 4.1 The k-error alignment method comprises three steps: Division of the query into $k + 1$ contiguous fragments (**a**); exact search for the fragments (**b**), and checking whether or not a k-error alignment has been found by filling in the dynamic programming matrix anchored by the exact match (**c**)

New Concept

Name	Comment
k-error alignment	speed up alignment by exact matching

New Programs

Name	Source	Help
kerror	book website	kerror -h
mutator	book website	mutator -h

Problem 235 Create a new directory for this section, KerrorAlignment, and change into it. Copy the sequence of dmAdhAdhdup.fasta, cut out positions 2301-2400, and save the resulting sequence fragment in dmAdhFrag.fasta. Check the correct region was cut out by matching the fragment back onto dmAdhAdhdup.fasta using keywordMatcher.

Problem 236 Mutate position 10 in dmAdhFrag.fasta using the program mutator and save the mutated sequence in dmAdhFrag2.fasta. Check the mutation appeared at the expected position using gal. Then compare the mutated fragment to the original full sequence to confirm the mutated fragment does not match exactly any more (keywordMatcher), but the inexact match remained intact (lal).

Problem 237 The program kerror implements k-error matching. Use it to locate the error-free fragment in the original full sequence. Then search for the mutated fragment. Print out the fragments into which the mutated query is divided (-L) and convince yourself only one of them has an exact match in S.

Problem 238 We have already seen global and local alignments. What type of alignment is generated by `kerror`?

Problem 239 Next, we try to locate the *Adh/Adh-dup* locus in the genome of *Drosophila melanogaster*. The genome of *D. melanogaster* consists of three autosomes, two sex chromosomes, and the mitochondrial genome. Their sequences are located in the following files:

Chromosome	Type	File
Left arm of chromosome 2	Autosome	dmChr2L.fasta
Right arm of chromosome 2	Autosome	dmChr2R.fasta
Left arm of chromosome 3	Autosome	dmChr3L.fasta
Right arm of chromosome 3	Autosome	dmChr3R.fasta
Chromosome 4	Autosome	dmChr4.fasta
Mitochondrial genome	Mitochondrial	dmChrMt.fasta
X chromosome	Sex chromosome	dmChrX.fasta
Y chromosome	Sex chromosome	dmChrY.fasta

Copy the genome files to your working directory and determine the length of each of these DNA sequences (`cchar`). How long is the genome of *D. melanogaster* in total?

Problem 240 The file `hamlet.fasta` contains Shakespeare's *Hamlet* in FASTA format. Copy it to your working directory and take a look at it (`less`) to find the opening sentence (Who's there?). How much longer is the genome of *D. melanogaster* than the tragedy?

Problem 241 Given that the genome files of *D. melanogaster* are large, we can save disk space by deleting them from our working directory

```
rm dmChr*.fasta
```

and creating the corresponding symbolic links:

```
ln -s ../Data/dmChr*.fasta .
```

Does the link have the same content as the original file (`diff`)? What is the size of a link compared to that of the original file (`ls -l`)?

Problem 242 Find *Adh/Adh-dup* in the genome of *D. melanogaster* using `kerror`. For this purpose, write a script that goes over the chromosomes and sets the number of errors

$$k = 1, 2, 5, 10, 20, 50, 100, 200, 500.$$

Where is *Adh/Adh-dup* located, and how many errors does the final alignment contain?

Problem 243 What is the number of errors per site at the *Adh/Adh-dup* locus?

Problem 244 Take another look at the *Adh/Adh-dup* alignment returned by `kerror`. Where are most of the errors located?

Problem 245 Instead of calculating the number of errors per site from a global/local alignment, calculate the number of mismatches per site from a local/local alignment. For this purpose, cut out the target region on chromosome 2 L and aligning it to `dmAdhAdhdup.fasta` using `lal`.

Problem 246 Repeat the `kerror` search for *Adh/Adh-dup* with the minimum *k* possible. Observe the run time printed by the program. How is this distributed between exact matching and checking through dynamic programming?

Problem 247 Align `dmGenomic.fasta` with its homologue from *D. guanche* using `gal`. What would *k* need to be for locating the *D. guanche Adh/Adh-dup* in the *D. melanogaster* genome? Is that feasible?

4.2 Fast Local Alignment

Homology between sequences is best determined locally: Even if homology were, in fact, global—or global/local as in *k*-error alignment—a local alignment would detect these configurations if they maximized the score. As we have seen for *k*-error alignment, local alignment is sped up by mixing exact and inexact matching. The first step in both methods, is therefore, to divide the query into shorter segments that are to be matched exactly. However, instead of using contiguous segments of variable length as in *k*-error alignment, in local alignment the segments, which are usually called "words", overlap and have a fixed length, say 11 bp (Fig. 4.2a). These words are then searched for in the subject sequence, which typically consists of an entire database of sequences (Fig. 4.2b). Wherever a hit is found, it is extended to the left and right until the score cannot be improved any further (Fig. 4.2c). In other words, the sketched algorithm returns *ungapped* local alignments. This was implemented in the first version of BLAST from 1990 [6]. Modern versions return gapped alignments [7], but as we shall see below, the ungapped algorithm is already quite effective.

Apart from its clever algorithm, BLAST is fast, because it incorporates a formula for calculating the significance of an alignment. The alternative to using the formula would be to compute *P*-values from simulations. As we show in this section, calculating *P*-values by simulation is much slower than by formula. Moreover, an important lesson to take away from this set of Problems is that small adjustments in parameters can have large effects on the alignments returned.

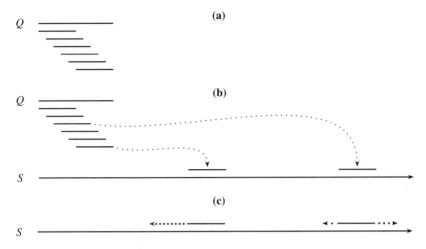

Fig. 4.2 BLAST algorithm. Preprocess the query sequence, Q, into overlapping words (**a**); search the words in the subject, S, (**b**); extend matches until the score cannot be improved any further (**c**)

New Concepts

Name	Comment
fast local alignment	ungapped or gapped version
ungapped alignment	alignment without dynamic programming

New Programs

Name	Source	Help
blastn	book website	blastn -h
sblast	book website	sblast -h

4.2.1 Simple BLAST

Problem 248 Create the directory `FastLocalAlignment` and change into it. Copy or link `dmAdhAdhdup.fasta` into your working directory. Cut out positions 3101-3200 from `dmAdhAdhdup.fasta`, save the fragment in `dmAdhFrag.fasta`, and align it to the original sequence using `sblast`. Output the word list generated by `sblast`. How many words does it contain? How long are these words?

Problem 249 Our query has length 100. What happens if it contains ten equally spaced mutations, for example at positions 1, 12,..., 100? Check your answer by writing a script that drives `mutator` to repeatedly mutate the fragment at the desired positions. Make sure the script is applied to a copy of the original fragment called, say, `dmAdhFrag2.fasta`, rather than to the original.

Problem 250 The word length applied by `sblast` can be set by the user. What is the range of values compatible with finding the mutated fragment `dmAdhFrag2.fasta`?

Problem 251 Fast alignment algorithms like BLAST are called "heuristic" in contrast to the "optimal", but slow, algorithms based on the dynamic programming. Can lal, which implements an optimal algorithm, align dmAdhFrag2.fasta?

Problem 252 Natural mutations are randomly distributed rather than equally spaced. Use mutator with mutation rates of 1, 2, 5, 10, 20, and 50 % to generate mutated versions of dmAdhFrag.fasta and search for them in dmAdhAdhdup.fasta with sblast. Iterate 100 times per mutation rate and plot the number of alignments found as a function of mutation rate. Make sure that only one alignment is counted per run. Hint: grep -A 1 includes the line *after* the match in the output.

Problem 253 Repeat the mutation analysis, only this time set the sblast threshold for acceptance of an alignment from its default value of 50 to 25 (-t). Plot the results for both thresholds in a single graph.

Problem 254 Use sblast to find the best local alignment between dmAdhAdhdup.fasta and dgAdhAdhdup.fasta. Then use lal to find the optimal local alignment. The programs sblast and lal use the same score scheme—which one finds the better alignment?

Problem 255 By default, sblast extends an alignment for up to 20 steps without improving the score before it gives up. What happens if you increase that number to 40 (-s) and rerun the the comparison between *D. melanogaster* and *D. guanche*?

Problem 256 Write a script called driveSblastDm.sh to drive the comparison between dmAdhAdhdup.fasta and all chromosomes of *D. melanogaster* using sblast. Hint: There is a special notation for iterating over all command line arguments; to test this, write the script example.sh

```
for a in $@
do
    echo ${a}
done
```

and run it as

```
bash example.sh *
```

What is the interval occupied by *Adh/Adh-dup*?

Problem 257 Use time to compare the run time of sblast and kerror when searching for dmAdhAdhdup.fasta in dmChr2L.fasta. Can you explain the run time difference between sblast and kerror?

Problem 258 Recall from Problem 247 that kerror could not locate dgAdhAdhdup.fasta in the genome of *D. melanogaster*. Try again with sblast. What are the coordinates of the homologous region?

4.2.2 Modern BLAST

Problem 259 Align the artificially mutated sequence dmAdhFrag2.fasta to dmAdhAdhdup.fasta using blastn with default parameters. Do you get a hit? What happens if you adjust the word size option (-word_size)?

Problem 260 What is the default word size of blastn? To answer this question, again start from the exact match in dmAdhFrag.fasta, copy it to dmAdhFrag2.fasta, and mutate it using mutate.sh. But this time vary the step length between mutations until you find the smallest step length that gives a hit. To make this more convenient, copy mutate.sh to mutate2.sh and change its first line from

```
for i in $(seq 1 11 100)
```

to

```
for i in $(seq 1 $1 100)
```

The $1 refers to the first argument on the command line, which means mutate2.sh can now be run as, say,

```
bash mutate2.sh 50
```

where 50 is the distance between mutations. In addition, insert

```
cp dmAdhFrag.fasta dmAdhFrag2.fasta
```

as the first line of mutate2.sh. To minimize the number of trials, we use a method called "binary search": Start with the smallest word size, 1, which gives no hit, and the largest, 100, which does. This establishes that the sought word length lies in the interval (1, 100). The trick is now, to repeatedly halve this interval. If the new midpoint results in a match, it becomes the right border of the interval to be halved next; conversely, if the new midpoint results in no match, it becomes the left border.

Problem 261 By default, blastn is run in a low-sensitivity mode called "megablast". This can be switched to a high-sensitivity mode by using

```
-task blastn
```

Repeat the binary search from before to find the default -word_size in this mode.

Problem 262 Compare the sensitivity of blastn in megablast and blastn mode by plotting the % alignments detected as a function of mutation rate, as in Problem 252. Again, start from the exact match in dmAdhFrag.fasta and mutate it using mutator. Use the tabular output format for this, -outfmt 7.

Problem 263 Regenerate dmAdhFrag2.fasta with mutation distance 11:

```
bash mutate2.sh 11
```

Table 4.1 Default score schemes of the local alignment programs used in this section

Parameter	sblast	blastn	lal
Match	1	2	1
Mismatch	−3	−3	−3
Gap opening	na	−5	−5
Gap extension	na	−2	−2

and run `blastn` in blastn mode and with appropriate word size like in Problem 259:

```
blastn -task blastn -word_size 10 -query dmAdhFrag2.
    fasta -subjectdmAdhAdhdup.fasta
```

to get the score line

```
Score = 141 bits (156), Expect = 8e-38
```

of which the simplest statistic is the raw score in round brackets, 156. It is computed using the score scheme printed at the end of the BLAST output:

```
Matrix: blastn matrix 2 -3
Gap Penalties: Existence: 5, Extension: 2
```

Under the conventions we have used so far, this means match is 2, mismatch −3, gap opening −5, and gap extension −2. Apart from the match score, these values are identical to those used by `lal`. This is shown in Table 4.1, which summarizes the score schemes for `sblast`, `blastn`, and `lal`. Repeat the `blastn` run with `lal` adjusted such that it returns the same raw score as `blastn`.

Problem 264 The score line begins with the bit score. This is intended to be more independent of the score scheme than the raw score. What happens to these two scores when you set match to 1 instead of the default 2 (`-reward`)?

Problem 265 Use `blastn` to align the *Adh/Adh-dup* region from *D. melanogaster* and *D. guanche*. Can `lal` find a better alignment under the same score scheme?

Problem 266 The amount of searching `blastn` does is determined among other things by the heuristic parameter `-xdrop_gap_final`. Its default is 100, higher values generally lead to a more thorough search. Try to find an `-xdrop_gap_final` value that returns the same alignment as `lal`.

Problem 267 We have tuned the parameter `-xdrop_gap_final` to get the same alignment as `lal`. Is it theoretically possible to find a better alignment with `blastn`, for example, by setting a different value for `-xdrop_gap_final`, without changing the score scheme?

Problem 268 Cut out the interval 3101–3200 from the *D. guanche Adh/Adh-dup* and save it in `dgAdhFrag.fasta`. Align the fragment with `dmAdhAdhdup.fasta` using `blastn`. Compare the `blastn` result with that obtained by `lal -A 2`. If the two results disagree, try reducing the `-word_size` from its default value of 11 to make the results agree.

Problem 269 The score line of our current BLAST command is

```
Score = 24.7 bits (26), Expect = 0.015
```

We have so far explored the bit score and the raw score. The `Expect` value means that 0.015 alignments between shuffled versions of the query and the subject are expected to have a score of 24.7 or greater. This value is called the expectation value or E-value. It is related to the significance from statistics through

$$P = 1 - e^{-E},$$

which implies $P \leq E$. What is P given $E = 0.015$?

Problem 270 The probability of finding an alignment with score, S, greater or equal to some given score, x, is [6]

$$P(S \geq x) = 1 - e^{-y},$$

where $y = Kse^{-\lambda x}$, s is the "effective" search space, K a correction factor, and λ incorporates the score matrix. These three parameters are quoted in the footer of the `blastn` output. Use their "gapped" version to compute $P(S \geq x)$ for our alignment.

Problem 271 To further explore the E-value computed by `blastn`, and hence the significance of the alignment, we test the null hypothesis that the observed score is due to chance through simulation rather than relying on the theory used by `blastn`. For this purpose, we compute the distribution of scores between random versions of `dmAdhAdhdup.fasta` and the original `dgAdhFrag.fasta`. Use `lal` with `blastn` parameters. Start with a small number of iterations until the simulation works, then carry out a long run with 1000 iterations and store the scores. How long does this take?

Problem 272 Write an AWK script called `count.awk` to compute the frequency of each distinct score. Plot this frequency as a function of the score.

Problem 273 The significance of the original score of 26, that is, its P-value, is the frequency, with which scores $\geq$ 26 appear among the random scores in `simPval.dat`. What is the simulated P-value of the original score? Compare your simulated result to the theoretical P-value implied by the BLAST output.

Problem 274 As a final step in our exploration of alignment statistics, we compare the theoretical score distribution to that generated by simulation in Problem 272. The probability of observing a score S is

$$P(S) = \lambda e^{w}, \tag{4.1}$$

where $w = (\mu - S)\lambda - e^{(\mu - S)\lambda}$ and $\mu = \ln(Ks)/\lambda$. The program `gnuplot` allows the definition of variables and functions, which can then be plotted; for example:

```
a=10
f(x)=a*x
plot f(x)
```

Use this feature to overlay the histogram from Problem 272 with the theoretical curve given by Eq. (4.1).

Problem 275 Write a script called `blast.sh` to search `dmAdhAdhdup.fasta` across all chromosomes of *D. melanogaster*. To minimize spurious results, use a minimum E-value of 10^{-20} by setting

```
-evalue 1e-20
```

Reduce clutter by switching to tabular format (`-outfmt 7`). How long does the search take?

Problem 276 Searches can be sped up by turning the subject sequences into a BLAST database:

```
cat dmChr*.fasta |
makeblastdb -dbtype nucl -title dmDb -out dmDb
```

How long does this take? What is the run time for searching `dmAdhAdhdup.fasta` in the database?

Problem 277 What happens to the run time when using the default megablast mode and the database? Does the mode affect the result?

Problem 278 Repeat the search with `dgAdhAdhdup.fasta`. Again compare the results obtained in blastn and megablast mode.

4.3   Shotgun Sequencing

The quest for ever faster alignment methods is driven by the ease with which genomes can be sequenced these days. When sequencing a DNA molecule, biologists initially need to know the sequence of about 20 nucleotides. This allows them to synthesize an oligonucleotide complementary to the known sequence. The oligo is then used to start, or *prime*, a sequencing reaction. Any such reaction reveals the identity of at most a few hundred nucleotides. A naïve way to sequence longer molecules would, therefore, be to obtain the first sequencing result, design a new primer to the 3' end of the sequence just determined, and repeat the cycle of sequencing and primer design. However, this sequential walking along a chromosome is slow. It is much more efficient to parallelize the procedure: Fragment a large number of copies of the template molecule through, for example, sonication. Then insert the random fragments into a piece of DNA with known sequence. The sequencing reaction can now be primed using always the same oligos complementary to the known flanking DNA. This "shotgun sequencing" was invented by the English biochemist Fred Sanger in

1982 [43] and yields random sequences, or *reads*, which need to be assembled into the template sequence. Assembly is carried out by programs that look for overlaps between a potentially very large number of reads.

In the following section, we start with optimal overlap alignments and end with a fast method for assembling the genome of *Mycoplasma genitalium* from simulated shotgun reads.

New Concepts

Name	Comment
overlap alignment	sequence assembly
shotgun sequencing	rapid sequencing method

New Programs

Name	Source	Help
oal	book website	oal -h
sequencer	book website	sequencer -h
velvetg	book website	velvetg
velveth	book website	velveth

Problem 279 The two sequences, S_1 = ACCGTTC and S_2 = GTTCAGTA overlap. Write down their alignment to show this.

Problem 280 Create the directory Shotgun, change into it and write S_1 and S_2 into the FASTA files s1.fasta and s2.fasta, respectively. Then use the program oal to compute the overlap alignment between S_1 and S_2.

Problem 281 Sequencing reads can originate from the forward or the reverse template strand. How many comparisons are necessary between two sequences to find the best overlap if you take strand into account?

Problem 282 The files f1.fasta, f2.fasta, and f3.fasta each contain a fragment of the *M. genitalium* genome. Copy them from the Data directory to your working directory. Think of these fragments as sequencing reads from a shotgun experiment. To find the overlaps between them, we need to compare each fragment to the forward and reverse strand of the other two fragments. If we denote the forward and reverse strands of the i-th fragment as f_i^f and f_i^r, the following combinations of fragments need to be checked:

	f_1^{f}	f_1^{r}	f_2^{f}	f_2^{r}	f_3^{f}	f_3^{r}
f_1^{f}			•	•	•	•
f_1^{r}						
f_2^{f}				•		•
f_2^{r}						
f_3^{f}						
f_3^{r}						

Think of the names along the first column as query, the names along the first row as subject. Write down the score for each comparison. What are the two most substantial overlaps and what does this say about the relationship between the three fragments? Hint: A read can be reverse complemented using revComp. Also, start by creating files with names like f1f.fasta and f1r.fasta.

Problem 283 Look at the overlap alignments of the two most substantial overlaps just discovered. Draw by hand a figure with alignment coordinates to show how the reads overlap. Then calculate the length of the genomic sequence from which the reads were taken.

Problem 284 Every sequencing project starts with the isolation of the template DNA molecule. We simulate this step by generating a random sequence. To give us an idea of how long a realistic genome might be, use cchar to compute the length of the genome of *M. genitalium* contained in the file mgGenome.fasta. Also, what is the fraction of G and C nucleotides in *M. genitalium*, that is, its GC content?

Problem 285 Use ranseq to generate a random genome with the same length and GC content as *M. genitalium*, and store it in ranGenome.fasta. To obtain exactly the same genome as we did, initialize the random number generator in ranseq with 35. Check your result with cchar.

Problem 286 What is the difference between ranseq and randomizeSeq?

Problem 287 A shotgun sequencing experiment yields a—usually large—number of random reads comprising a total of s nucleotides. Such an experiment is characterized by the coverage, $c = s/L$, where L is the length of the molecule sequenced. How many bases would you get if the sequencing facility at your institution sequenced ranGenome.fasta to a coverage of 10?

Problem 288 We now study the relationship between the coverage, c, and the probability of sequencing a particular nucleotide. If we picture "sequencing" as randomly drawing a single nucleotide from the genome, the probability of getting a particular one is $1/L$ and the probability of not getting it is $1 - 1/L$. The probability of not sequencing a nucleotide in s trials is

$$P_0 = \left(1 - \frac{1}{L}\right)^s.$$

To simplify this, we first rewrite

$$\left(1 - \frac{1}{L}\right)^s = e^{s \ln(1 - \frac{1}{L})},$$

and use the approximation $\ln(1 + x) \approx x$ to get

$$P_0 \approx e^{-s/L} = e^{-c}.$$

How many nucleotides are expected to be left unsequencedif the random genome is shotgunned to a coverage of 10?

Problem 289 What is the theoretical coverage necessary to achieve a combined gap length of 1?

Problem 290 What is the theoretical coverage necessary to achieve a combined gap length of 0?

Problem 291 Use `sequencer` to sequence your random genome to the coverage that would yield an expected gap length of 1. Store the reads in `reads.fasta`. How many reads did you generate? How many nucleotides?

Problem 292 We use `velvet` for assembly. `Velvet` first "hashes" the reads, which means indexing overlapping read fragments of some length. These fragments are then used as seeds for inexact matches between reads. Use `velveth` to hash the short reads to a length of 21 and `Assem` as the name of your assembly directory.

Problem 293 Assemble the hashed reads using `velvetg`. The only option you need to set at this stage is the expected coverage, `-exp_cov`. `velvetg` assembles the reads into contiguous sequences, or *contigs*. How many contigs do you get and how many nucleotides do they comprise?

Problem 294 So far we have simulated sequencing projects by picking random reads. In real sequencing projects, random fragments of mean length, say, 500 bp are picked and sequenced from both sides. This approach is called *paired-end sequencing*. The information about read pairing is passed on to the assembly program. Repeat the sequencing experiment, but this time generate paired-end reads in `sequencer` and use the corresponding option in `velveth`. In `velvetg` set the "insert length", that is, the length of the fragment sequenced from both ends, to 500. How many contigs and nucleotides do you get?

Problem 295 You have sequenced with the default error rate of 0.1%, which means the sequencer mis-calls one in 1000 nucleotides. What happens if you eliminate errors altogether?

Problem 296 Increase the error rate to 1%. How many contigs and nucleotides do you get?

Problem 297 Sequence and assemble the genome of *M. genitalium*. Generate single-end reads in `sequencer` with default error and the coverage for an expected gap length of 1 calculated in Problem 289. How many nucleotides do you get in how many contigs?

Problem 298 So far, we have assessed the quality of alignments only by counting the contigs and the nucleotides they contain. However, a popular measure of assembly quality is the N_{50}. This is related to the median contig length, which in turn is similar to the mean contig length. So before, we define the N_{50}, we remind ourselves of median and mean. Consider five toy contigs with lengths $\mathscr{L} = \{2, 2, 3, 4, 5\}$. What is the median and the mean contig length?

Problem 299 The N_{50} is the length of the shortest contig contained in the set of longest contigs that comprise at least 50% of assembled nucleotides. What is the N_{50} of our toy contigs, $\mathscr{L}$?

Problem 300 On its last output line, `velvetg` also prints the N_{50} (n50), because the larger the N_{50}, the better the assembly. What is the N_{50} of the assembly you have just computed?

Problem 301 We now write a pipeline to compute the N_{50} for the set of contigs assembled in the previous Problem. First, we write a pipeline for extracting and sorting the contig lengths. It should, executed from left to right,

- report the length of each contig (`cchar -s`);
- extract the lines that contain the contig length (`grep`);
- print only the contig length (`awk`);
- sort the lengths in descending order (`sort`).

What is the median contig length (`tail` & `head`)?

Problem 302 Now we are in a position to write an AWK program for computing the N_{50} from ordered contig lengths. Call this program `n50.awk`; it

- stores the lengths of all contigs;
- extracts the total length of all contigs;
- sums the contig lengths, and once that sum is greater than half the total length, prints the current contig length. Use a `while` loop for this step.

Problem 303 Repeat the sequencing and assembly with paired-end reads and compute the N_{50}. Save this final assembly to `mgAssembly.fasta`.

4.4 Fast Global Alignment

In Sect. 4.3 we assessed the quality of our genome assembly using such metrics as the number of contigs, the number of bases assembled, and the N_{50}. However, the ultimate quality check is to align the assembly and the template. Since these two data sets comprise over 500 kb, the corresponding alignment matrix would be huge. Instead, we can apply fast, "heuristic" alignment methods that are based on a clever mix of exact matching and optimal alignment. In this section, we use the package MUMmer [15, 29] for fast global alignment. Specifically, we assess the assembly of the *M. genitalium* genome and to compare two strains of *Escherichia coli*. The core program in the MUMmer package, mummer, is based on a suffix tree. In Sect. 3.2 the suffix tree was constructed from a single sequence. This can be generalized to an arbitrary number of sequences by adding the suffixes of them all. However, we now need to keep track of where a suffix came from and hence label the leaves with a pair of numbers: (string, suffix). For example, Fig. 4.3a shows the suffix tree for ADAM already familiar from Fig. 3.5. We first add a sentinel character (Fig. 4.3b) and then augment the leaf labels by information on string of origin (Fig. 4.3c). Finally, we insert the suffixes of DAD (Fig. 4.3d). A leaf can now have more than one label; for example (1, 5) and (2, 4) for the sentinal suffix, $. In this section, we first learn how generalized suffix trees can be used to find repeats between genomes. Then, we use MUMmer to compare bacterial genomes.

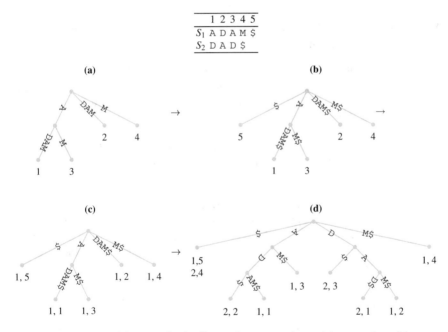

Fig. 4.3 Construction of the generalized suffix tree for ADAM and DAD. (**a**): ADAM alone; (**b**): ADAM plus sentinel ($); (**c**): ADAM$ with leaf labels of the form (sequence, suffix); (**d**): ADAM$ and DAD$ in one suffix tree

New Concept

Name	Comment
generalized suffix tree	suffix tree for ≥ 1 sequences

New Programs

Name	Source	Help
drawStrees	book website	drawStrees -h
mummer	book website	mummer -h
nucmer	book website	nucmer -h
show-snps	book website	show-snps -h

Problem 304 Draw by hand the generalized suffix tree of $S_1 =$ AAGCG and $S_2 =$ AGT.

Problem 305 Use the program drawStrees to draw the generalized suffix tree of S_1 and S_2. drawStrees generates a LaTeX file as output. Convert this to a viewable postscript file as described in Problem 185.

Make sure that in the FASTA input for drawStrees the second sequence is terminated by a carriage return.

Problem 306 Label by hand each internal node in your suffix tree with the length of the path label, the string depth. By searching for the node with the greatest string depth you can find the longest repeat in the input data. What is the longest repeat between S_1 and S_2?

Problem 307 Determine the length and composition of the assembly in mgAssembly.fasta generated in Problem 303 and compare this to the length of the original genome.

Problem 308 Before we compare mgAssembly.fasta to mgGenome.fasta, we carry out a few simple experiments with mummer to help interpret its output. Use ranseq to generate a random 1 kb sequence and save it in s1.fasta. Then use cutSeq to cut the first and last 100 bp from s1.fasta, splice them together, and save the resulting single 200 bp sequence to s2.fasta. Compare s1.fasta and s2.fasta using mummer. Can you interpret the output?

Problem 309 In its simplest form, the output of mummer is a dot plot, albeit a very efficiently computed dot plot. Use the program mum2plot.awk to convert the mummer output to gnuplot input.

```
mummer s1.fasta s2.fasta | awk -f mum2plot.awk |
    gnuplot ...
```

In the resulting plot, which sequence is written along the x-axis, which along the y-axis?

Problem 310 Reverse-complement `s2.fasta` using `revComp`, save it to `s3.fasta`, and align it again to `s1.fasta`; only this time we require that `mummer` returns matches on both the forward and reverse strand (`-b`) and that these matches are reported relative to the original sequence, rather than the reverse complement (`-c`). Plot the result as before.

Problem 311 As a final preparatory step, write the contents of `s2.fasta` and `s3.fasta` to `s4.fasta` and compare it to `s1.fasta`. Before carrying out the computation, make a sketch of the plot you expect so see.

Problem 312 When comparing the output of `mgAssembly.fasta` to `mgGenome.fasta`, it is easier to interpret the plot if the assembly consists of a single sequence. To concatenate the contigs into a single sequence, first create a new file with a FASTA header

```
echo '>Assembly' > mgAssemblyS.fasta
```

Then append the contigs with their headers removed:

```
sed '/^>/d' mgAssembly.fasta >> mgAssemblyS.fasta
```

Compare `mgGenome.fasta` as reference to `mgAssemblyS.fasta` as query using `mummer`. As before, it should compute matches on both the forward and the reverse strand, and report all query positions with respect to its forward strand. Hint: `mummer -help`.

Problem 313 Our next goal is to compare the genomes of two assembled *E. coli* strains contained in `ecoliK12.fasta` and `ecoliO157H7.fasta`. How long are these two genomes?

Problem 314 Use `mummer` to draw the dot plot for K12 and O15H7 with K12 as reference. What do you notice about the two sequences?

Problem 315 Alignments are used to compare sequences at nucleotide resolution, something that cannot be done with dot plots. Resolution at the nucleotide level can, in turn, be used to estimate the number of mutations, or single nucleotide polymorphisms (SNPs), that separate two genomes. The program `nucmer`, which is part of the mummer package, aligns genomes ready for SNP discovery by `show-snps`:

```
nucmer -p nucmer ecoliK12.fasta ecoliO157H7.fasta
show-snps nucmer.delta > nucmer.snps
```

What is the number of SNPs per nucleotide between K12 and O157H7?

Problem 316 Given that the mutation rate in *E. coli* is 2.2×10^{-10} per generation [31], what is the divergence time between K12 and O157H7 in generations?

Problem 317 The doubling times of *E. coli* range from 30 to 90 min [39]. What is the corresponding range in years to the most recent common ancestor for K12 and O157H7?

4.5   Read Mapping

Sequencing reads are among the most common data in Molecular Biology. We have already seen one typical application of sequencing, shotgun sequencing, where reads are assembled into the template sequence. In contrast to this *de novo* sequencing, re-sequencing is carried out in organisms with established reference genome sequences. In this situations, reads are mapped to the reference, often for discovering genetic variants. In this section, we explore the time requirements of read mapping, and how read alignments can be manipulated.

New Concept

Name	Comment
read mapping	variant discovery

New Programs

Name	Source	Help
bwa	book website	bwa
samtools	book website	samtools

Problem 318 Create the directory ReadMapping for this session and change into it. Use the program sequencer to simulate the sequencing of dmChr2L.fasta with default parameters, except for coverage, which should be 15 instead of 1. In addition, you can synchronize your results with ours by using 10 as the seed for the random number generator (-s). Save the reads in reads.fasta. How many reads were generated (grep)? Do they represent the desired coverage (cchar)?

Problem 319 The program sequencer picks a fragment of mean length 500 bp with standard deviation $\sqrt{500} \approx 22.4$ (see sequencer -h). From such a fragment it generates a read of length 100 bp. Modern sequencers produce reads of identical lengths and hence, the standard deviation of read length is zero by default. However, occasionally sequencer produces a read shorter than the default length. How many of these do you find in your data set? Can you explain how they might occur?

Problem 320 Map the reads using the BLAST-mode blastn-short for mapping short reads (-task). But before mapping all reads, write a script for measuring the run times when mapping 1, 2, 5, 10, 20, 50, 100, 200, 500, and 1000 reads. How long would it take to map all reads?

Problem 321 The program bwa is a read mapper based on the Burrows–Wheeler transform, hence the name [33]. It works on an index of the reference sequence, which is computed with the command

```
bwa index -p dmChr2L dmChr2L.fasta
```

How long does index computation take? The program reports the main computational steps. Do you recognize any of the operations being described?

Table 4.2 Mandatory fields in SAM file

Col.	Meaning
1	Query
2	Comment flag; 0: none, 4: unmapped, 16: reverse-complement
3	Subject
4	Position
5	Mapping quality
6	Match string; M: match, D: deletion, I: insertion, S: soft clipping from read
7	Name of read mate
8	Position of read mate
9	Template length
10	Read sequence
11	Base quality

Problem 322 Like with BLAST in Problem 320, align 1, 2, 5, 10, 20, 50, 100, 200, 500, and 1000 reads and redirect the output to the null device. But unlike the BLAST run, these are *thousands*, which requires changing the assignment to x in line 4 of runBlast.sh. Call the new script runBwa.sh. Plot the run times and estimate how long it would take to align all reads.

Problem 323 Align all reads and save the result in reads.sam. How long does this take? Compare the observed run time to the estimate from Problem 322.

Problem 324 Take a look at reads.sam (head). Lines starting with @ are header lines. The body of the file contains eleven mandatory columns, which are listed in Table 4.2. Column 6 contains a description of the alignment like 100M, which means 100 matches, in other words, an exact match. Find such an exact match and use keywordMatcher to locate it in dmChr2L.fasta. How are reverse-complemented reads represented in reads.sam?

Problem 325 The first things we might want to do with a SAM file is to look at the alignments. This is done using the program samtools. To prepare reads.sam for viewing, convert it first to its binary format and sort it:

```
samtools view -b reads.sam | samtools sort > reads.bam
```

These commands can be sped up by assigning more than one thread to each process (-@). Next, the BAM file is indexed

```
samtools index reads.bam
```

and can be viewed

```
samtools tview -reference dmChr2L.fasta reads.bam
```

Can you explain the commas and dots you see?

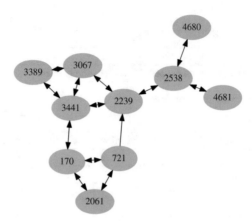

Fig. 4.4 Example of a protein family defined by BLAST hits. Nodes represent proteins, edges are either reciprocal hits of the form query ↔ subject, or unidirectional hits between query → subject

Problem 326 The help menu of `tview` is switched on by pressing "?". What is the command for visiting a particular position in the alignment?

Problem 327 Positions in an alignment are denoted by

```
sequenceName:position
```

The names of the template sequences mapped to are listed in the header of a SAM/BAM file:

```
samtools view -H reads.bam
```

where `SN` in the output refers to "sequence name". Go to position 2000. Given a position annotation like

```
2001
ATTC
```

which digit refers to the actual position?

4.6   Clustering Protein Sequences

In this section, we use protein BLAST to find out whether the genome of *M. genitalium* contains sets of genes with similar function. We do this by searching for sets of genes with similar sequences and assume tacitly that sequence similarity implies similar function. Such sets of similar genes are called "gene families". Gene families arise in the course of evolution by gene duplication [35], as seen with *Adh/Adh-dup* in *Drosophila*. A gene family might look like the graph in Fig. 4.4, where the nodes represent proteins and the edges BLAST hits, i.e., the arrows pointing from query to subject.

In homologous pairs, either of the two proteins involved can serve as query/subject. If there is a hit with both labelings, it is called "reciprocal" and is denoted as query $\leftrightarrow$ subject. Otherwise we just have a hit in one direction, query $\rightarrow$ subject. Reciprocal hits are the norm. The one exception in Fig. 4.4 is the pair $721 \rightarrow 2239$, where the comparison is only significant if 721 is query, not if 2239 is query.

Gene families are common in the large genomes of, for example, mammals. The human genome encodes approximately 23,000 genes of which hundreds belong to several families of olfactory receptors [11], a reflection of the importance of the sense of smell in terrestrial mammals. However, *M. genitalium* has one of the smallest genomes of any free living organism with less than 500 protein-coding genes. The question we investigate in this section, is therefore, whether the genome of *M. genitalium* is too small to accommodate gene families.

New Concept

Name	Comment
protein family	set of proteins with similar sequences and functions

New Programs

Name	Source	Help
blast2dot.awk	book website	blast2dot.awk -v h=1
blastp	book website	blastp -help
circo (GraphViz)	package manager	man circo
getSeq	book website	getSeq -h
neato (GraphViz)	package manager	man neato
ranDot.awk	book website	ranDot.awk -v h=1

Problem 328 How many edges could in theory be drawn between the nodes (proteins) in Fig. 4.4? What proportion of these is actually found?

Problem 329 Make a directory for this session,

FastLocalAlignmentProt

change into it, and link the file mgProteome.fasta, which contains all proteins of *M. genitalium*. How many proteins are there?

Problem 330 In order to find protein families in *M. genitalium*, we shall carry out an all-against-all comparison of the proteins. How many comparisons does this comprise? Test your prediction by producing five identical DNA sequences using ranseq with the same seed for the random number generator, and running blastn on that toy data set.

Problem 331 Carry out an "all-against-all" comparison on mgProteome.fasta with $E = 10^{-5}$. Save the result in tabular format in the file mgProteome.blast. Since we are only interested in the closest homologue of each sequence, restrict the output to at most one hit per query:

```
-max_hsps 1
```

How long does it take `blastp` to carry out the approximately 230,000 comparisons (`time`)? Take a look at the BLAST result.

Problem 332 Convert `mgProteome.fasta` to the BLAST database `mgProteome` (`makeblastdb`). How long does this take? How long does the all-against-all comparison take using the database? Again, save the result in `mgProteome.blast`.

Problem 333 Every accession in `mgProteome.blast` has the redundant prefix `lcl|`. Use `sed` to remove it. Hint: Save the result of `sed` to a temporary file, say `tmp`, before replacing the original with the edited version.

Problem 334 What happens if the input file of a command is also the output, without saving to an intermediary file? For example

```
sed expr mgProteome.blast > mgProteome.blast
```

Try this, but only after making a backup copy of `mgProteome.blast` first.

Problem 335 Look at the first ten lines of `mgProteome.blast`. Can you find a protein with at least one homologue in the proteome (other than itself)? Use `grep` on `mgProteome.fasta` to discover the function of the first such protein in the list and of its homologue.

Problem 336 How many lines are contained in `mgProteome.blast`? Is it sensible to analyze this file manually to find protein families?

Problem 337 What is the protein with the largest number of homologues in the proteome of *M. genitalium* (`sort`, `uniq -c`)?

Problem 338 Write a pipeline that prints out the list of unique MG-numbers of the proteins linked to MG_410 by BLAST hits. Save the names of this protein family in `protFam.txt`.

Problem 339 The program `neato` is used for visualizing large graphs. We plan to use it for visualizing the network of BLAST hits in the proteome of *M. genitalium*. Neato takes input written in the dot language. Here is an example:

```
graph G {
    a -- b
    b -- c
    c -- a
}
```

If this is contained in `example1.dot`, `neato` calculates the layout and visualizes it

```
neato -T x11 example1.dot
```

If the x11-terminal is not available on your system, try generating postscript output

```
neato -T ps example1.dot > example1.ps
```

which can then be viewed

```
gv example1.ps &
```

to give

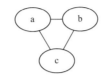

Write a file `example2.dot` that specifies the following graph:

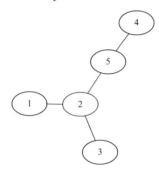

Problem 340 To represent reciprocal and unidirectional BLAST hits, use

```
a -- b [dir=both]      # reciprocal hit
c -- d [dir=forward]   # unidirectional hit
```

which sets the edge attribute `dir`, the default value of which is none. Write the dot code for

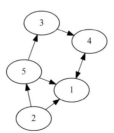

Problem 341 Use the program `ranDot.awk` to draw a random graph with ten nodes, directed edges, and an edge probability of 0.5:

```
awk -f ranDot.awk -v n=10 -v p=0.5 -v dir=1 > ranDot.
    dot
```

What is the expected number of edges in this random graph? Compute the observed number of edges, counting reciprocal edges double. Compare the expected number of edges to that observed.

Problem 342 Visualize `ranDot.dot` using `neato`. Then visualize it with the program `circo`, which has the same syntax as `neato` and is also part of the Graphviz package. Which graph do you prefer?

Problem 343 We now convert the hits in `mgProteome.blast` to dot code:

```
awk -f blast2dot.awk mgProteome.blast |
sed 's/MG_//g' > mgProteome.dot
```

Notice the `sed` command for removing `MG_` to save space when drawing the nodes. Choose between `neato` and `circo` to draw the graph. What is the size of the largest protein family? Compare it to `protFam.txt`, and if there are any differences, write the new version of the protein family to `protFam2.txt`.

Problem 344 By default, `blast2dot.awk` only includes proteins with at least one BLAST hit in the output. To include proteins without homologues, repeat the computation of dot code including the "singletons"

```
awk -f blast2dot.awk -v singletons=1 mgProteome.blast|
sed's/MG_//g' > mgProteome2.dot
```

What proportion of the proteome are singletons?

Problem 345 What is the function of the members of the extended protein family? Find out by looking up the annotations contained in their FASTA headers using `grep` with extended regular expression syntax, which allows OR. For example,

```
grep -E '(a|b)'
```

reports lines matching a or b. Hint: Use `protFam2.txt` to automatically construct the list of alternative matches.

Problem 346 To learn more about the protein family just obtained, we search a database of protein motifs and annotations. Prosite is one such database. It consists of two files, `prosite.dat` and `prosite.doc`. The file `prosite.dat` contains motifs, `prosite.doc` annotations. Search `prosite.doc` for our gene family. What is its function?

Problem 347 Why should a bacterium possess multiple ABC transporters?

4.7   Position-Specific Iterated BLAST

For all the alignments computed so far, we used the same score scheme at every position; any mutation between, say, a proline and a histidine had the same score, regardless of whether it occurred in an active site, a transmembrane helix, or an extracellular loop. An alternative to the traditional position-invariant score matrices such as PAM70 and BLOSUM62, are position-specific score matrices; here is an example:

```
        A   R   N   D   C   Q   E   G   H   I   L   K   M   F   P   S   T   W   Y   V
  1 M  -1  -1  -2  -3  -1   0  -2  -3  -2   1   2  -1   5   0  -2  -1  -1  -1  -1   1
  2 E  -1   0   0   1  -4   2   5  -2   0  -3  -3   1  -2  -3  -1   0  -1  -3  -2  -2
  3 K  -1   2   0  -1  -3   1   1  -1  -1  -3  -2   4  -1  -3  -1   0  -1  -3  -2  -2
...
329 N  -2   0   6   1  -3   0   0   0   1  -3  -4   0  -2  -3  -2   1   0  -4  -2  -3
```

At every position the query amino acid is listed next to the 20 possible scores. Such a matrix has the advantage of reflecting the local variation in mutation probabilities. Its disadvantage is that every query/subject combination requires a new score matrix. Therefore, position-dependent score matrices cannot be precomputed; they need to be constructed on the fly through iteration. In the first iteration, a conventional score matrix is used to find the closest homologues of the query, which are converted to an initial position-dependent score matrix. This is applied in the second round of sequence comparison. Since searches based on position-dependent score matrices are usually more sensitive than traditional searches, the second iteration often uncovers new homologues. These are used to revise the score matrix before repeating the search. The cycle of searching and matrix revision continues until no new sequences are found. The BLAST program `psiblast`, for Position-Specific Iterated BLAST [7], implements such an iterative search. In this section, we use it to search for further members of the ABC transporter gene family in the proteome of *M. genitalium*.

New Concept

Name	Comment
position-specific iterated BLAST	more sensitive than regular BLAST

New Program

Name	Source	Help
psiblast	book website	psiblast -h

Problem 348 The program `psiblast` takes as input a query sequence and a database in BLAST format. Set up the directory `PsiBlast` for this session. Then convert `mgProteome.fasta` to the BLAST database `mgProteome`. We previously found that MG_410, together with MG_180 and MG_179, has the largest number of BLAST hits in the proteome of *M. genitalium*. So save MG_410 to `mg_410.fasta` and take it as the starting sequence for our initial `psiblast` run. Leave all options unchanged except for the output format, which should be tabular. Save the output in `mg_410.psi`. How many distinct hits have an E-value less than 10^{-5}?

Problem 349 Repeat the comparison of `mg_410.fasta` to the full proteome with `blastp`. Save the result in `mg_410.bp` and compare it to `mg_410.psi`. Can you find any differences?

Problem 350 Rerun `psiblast` iteratively until convergence by using

`-num_iterations 0`

Save the output in `mg_410b.psi`. How many rounds does `psiblast` go through? How many distinct proteins have an E-value of 10^{-5} or smaller in the last round? Save the corresponding proteins in `psiBlastList.txt`. Hint: Each round starts with the best hit, MG_410 to itself.

Problem 351 The list of ABC transporter proteins extracted from the graph of BLAST hits in *M. genitalium* comprised 18 sequences and was recorded in `protFam2.txt`:

```
MG_014 MG_015 MG_042 MG_065 MG_079 MG_080 MG_119 MG_179 MG_180
MG_187 MG_290 MG_303 MG_304 MG_410 MG_421 MG_467 MG_526 MG_390
```

What are the extra three proteins just identified?

Problem 352 What are the annotations in `mgProteome.fasta` of the extra proteins just found?

Problem 353 Next, we investigate the position-specific score matrix used by `psiblast`. To generate it, rerun `psiblast` and save the score matrix in `psiBlast.mat` using

`-out_ascii_pssm psiBlast.mat`

Which part of `psiBlast.mat` contains the position-specific score matrix? How many lines does this matrix consist of? Compare the length of the matrix to the length of MG_410.

Problem 354 Scores for a particular amino acid can vary widely along a protein sequence. To illustrate this, first identify the most frequent amino acid in MG_410 (`cchar`). Then use AWK to extract the scores for each position occupied by that amino acid from `psiBlast.mat`. What is the range of its match scores? Compare that to the corresponding match score in BLOSUM62.

4.8 Multiple Sequence Alignment

Up to now we have looked only at alignments between two sequences. But one often needs to align more than two sequences. To do this optimally for n sequences, an n-dimensional dynamic programming matrix is needed. Figure 4.5 illustrates such multidimensional matrices starting from zero dimensions, a mere dot in Fig. 4.5a. By doubling this dot and drawing a connecting edge, a one-dimensional matrix is

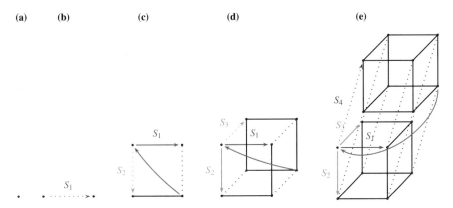

Fig. 4.5 Building multidimensional dynamic programming matrices for optimal multiple sequence alignment. The number of dimensions ranges from zero (**a**) to four (**e**) and corresponds to the number of sequences, S_i, that can be written along its edges and hence aligned. Sequences are indicated by colored arrows labeled S_i. They all start at the same node. The red arc indicates the trace-back

generated, which can accommodate a single sequence, S_1, written from left to right as indicated by the arrow in Fig. 4.5b. In the next doubling step we get the familiar two-dimensional matrix for aligning two sequences, S_1 and S_2, in Fig. 4.5c. The trace-back is indicated by the red arc. Repeat the doubling to get a cube for aligning three sequences (Fig. 4.5d). This can be repeated to create the four-dimensional hyper-cube in Fig. 4.5e.

Clearly, the space and time required for aligning by dynamic programming is multiplicative in the lengths of the input sequences. As a result, there are only few programs implementing optimal multiple sequence alignment [20, 34]. In practice, "shortcuts", or "heuristics", are taken. The most important of these is to reduce a multiple sequence alignment to a set of pairwise comparisons. In this section, we use two versions of this heuristics, query-anchored alignment, and progressive alignment to analyze sets of hemoglobin sequences.

New Concepts

Name	Comment
optimal multiple sequence alignment	generalize optimal pairwise alignment
progressive multiple sequence alignment	heuristic multiple sequence alignment method
query-anchored alignment	heuristic multiple sequence alignment method

New Programs

Name	Source	Help
`clustalw` or `clustal2`	book website	`clustal[w2] -help`
`fasta2tab.awk`	book website	`fasta2tab.awk -v h=1`
`shuffle.awk`	book website	`shuffle.awk -v h=1`

4.8.1 Query-Anchored Alignment

Problem 355 Set up the directory `Msa` for this session and obtain the following four hemoglobin sequences from Uniprot/Swissprot collection of curated protein sequences. This is contained in `uniprot_sprot.fasta`, from where individual sequences can be extracted using `getSeq`:

Accession	Protein
HBA_HUMAN	Human α-hemoglobin
HBA_HORSE	Horse α-hemoglobin
HBB_HUMAN	Human β-hemoglobin
HBB_HORSE	Horse β-hemoglobin

Save the sequences in files called `hbaHuman.fasta`, `hbaHorse.fasta`, etc.

Problem 356 The program `blastp` has an output format for anchoring all subject sequences to the query.

```
-outfmt 2
```

Use this with human α-hemoglobin as query and the other three as subject. Begin by constructing a file containing the subject sequences. The output contains amino acids printed like this

```
\
|
G
```

Can you guess what this means?

Problem 357 Repeat the query-anchored alignment, only this time use human β-hemoglobin as the subject.

4.8.2 Progressive Alignment

Problem 358 Query-anchored alignments have the disadvantage that they depend on the query. The progressive alignment algorithm, which is the most widely used method for computing multiple sequence alignments, avoids this. Given a set of sequences (Fig. 4.6a), progressive alignment starts by computing their pairwise alignments and hence distances (Fig. 4.6b). The distances are summarized as a tree (Fig. 4.6c), which groups the pair with the smallest distance first. At each internal node in the tree two clusters are joined. The tree is traversed from the leaves toward the root, and at the first internal node encountered the sequences of the corresponding leaves, A and D, are aligned (Fig. 4.6d). At the next internal node up, the pair B, C is aligned (Fig. 4.6e). At the root, finally, the two pairwise alignments generated so far, (A, D) and (B, C), are each treated as a unit and aligned with each other in the final pairwise alignment (Fig. 4.6f). In this way, as the algorithm progresses toward the root, only pairwise alignment problems need to be solved, which greatly simplifies the task of aligning multiple sequences. The disadvantage of this procedure is that any gaps introduced early in the algorithm cannot be changed later. This is why the tree is traversed from the leaves up rather than from the root down; the most similar sequences are aligned first, and their alignments contain the fewest of those irrevocable gaps.

We now recapitulate these steps using our four example sequences. To begin with, compute the $\binom{4}{2} = 6$ pairwise scores using gal together with the BLOSUM62 matrix. To convert a given score, s_{ij}, to a distance, d_{ij}, let s_m be the maximum score; then we compute the distance by dividing by the maximum and taking the complement:

$$d_{ij} = 1 - s_{ij}/s_m.$$

Problem 359 From the distance matrix, construct the order of clustering in the form of a guide tree as shown in Fig. 4.6b.

Problem 360 The hemoglobin proteins are homologs. In previous sections about *Adh* and its duplicate *Adh-dup* in *Drosophila*, we distinguished between two types of homologs: Orthologs, which have diverged through speciation, and paralogs, which have diverged through gene duplication. Classify our example genes into paralogs and homologs.

Problem 361 Save the four hemoglobin sequences in hemoglobin.fasta and align them using clustalw. Take a look at hemoglobin.dnd, which contains the guide tree in text format. To understand the relationship between a tree as text and a tree as graph, consider the guide tree in Fig. 4.6c. Its textual representation is

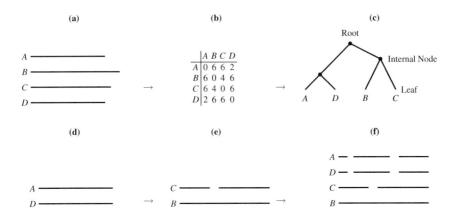

Fig. 4.6 Alignment of multiple sequences as progressive pairwise alignment along a guide tree: Four sequences (**a**) are aligned pairwise and their distances stored in a matrix (**b**), which is summarized as tree (**c**); traversal of this tree from the leaves to the root guides the alignment (**d–f**)

```
((A:0.5,D:0.5):1,(B:1,C:1):0.5);
```

This notation means

Token	Explanation
(	cluster opened, draw a node that is either the root or an internal node
A	leaf with label *A*
:0.5	distance to parent node is 0.5
,	separate nodes
)	cluster closed
;	end of tree description

The program `clustalw` uses unrooted guide trees. Without a root our example tree would be denoted as

```
((A:0.5,D:0.5):1.5,(B:1,C:1));
```

and look like this

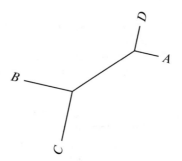

Draw by hand the guide tree produced by `clustalw`.

Problem 362 Take a look at the alignment in `hemoglobin.aln`. How many gaps does it contain, and during which phase of the algorithm was each of them introduced?

Problem 363 Let us examine the N-terminus of the alignment:

```
HBB_HORSE        -VQLS
HBB_HUMAN        MVHLT
HBA_HORSE        -MVLS
HBA_HUMAN        -MVLS
```

This contains one monomorphic site, L. However, the following arrangement of gaps gives two monomorphic sites, L and V:

```
HBB_HORSE        -VQLS
HBB_HUMAN        MVHLT
HBA_HORSE        MV-LS
HBA_HUMAN        MV-LS
```

Should not it be better than the first alignment? This depends on the treatment of gap-opening. By default, `clustalw` scores the opening of end gaps as zero. Consider the two sequences $S_1 = ATG$ and $S_2 = AG$. If gap opening is scored the same at every position, the optimal alignment should be

```
ATG
A-G
```

What does the `clustalw` alignment of these two sequences look like?

Problem 364 The gap-opening penalty of `clustalw` is set using

```
-GAPOPEN=x
```

Can you find a value for `x` that is compatible with the alternative N-terminal alignment:

```
HBB_HORSE       -VQLS...
HBB_HUMAN       MVHLT...
HBA_HORSE       MV-LS...
HBA_HUMAN       MV-LS...
```

Apply this to the hemoglobin sample.

Problem 365 To investigate the run time of `clustalw`, apply it to a single pair of random sequences of lengths 100, 200, 500, 1000, 2000, 5000, 10000, and 200000 bp. Plot the run times as a function of sequence length. How does the run time scale with sequence length?

Problem 366 Repeat the run time analysis, except this time vary the number of sequences aligned rather than their lengths. Generate 2, 5, 10, 20, 50, 100 random sequences of length 1 kb and measure the run times of `clustalw`. How does the run time scale with sample size?

Problem 367 Extract a random sample of 100 hemoglobin sequences from Uniprot/Swissprot (`uniprot_sprot.fasta`). For this, get the sequences with `getSeq` and convert them to tabular format by piping them through `fasta2tab`. `awk`. Shuffle the sequences using `shuffle.awk` to get an unbiased sample, then convert them back to FASTA format (`tr` and `fold`). How long does `clustalw` take to analyze this sample of 100 sequences? Repeat the time measurement with 200 random hemoglobin sequences.

Problem 368 How many hemoglobin sequences are contained in Uniprot? Estimate the time it would take to align all of them with `clustalw`.

Problem 369 Test your run time prediction.

Chapter 5
Evolution Between Species: Phylogeny

5.1 Trees of Life

In 1834 Charles Darwin (1809–1882) wrote in his notebook "I think"—but instead of finishing the sentence, he drew a phylogeny. When he published his thinking in 1859, *The Origin of Species* contained a single figure: a phylogeny. We already saw in Sect. 4.8 that phylogenies in the form of guide trees are useful for computing multiple sequence alignments. But beyond clever computing, phylogenetic trees embody biologists' thinking. In this section, we describe how to write, visualize, and traverse trees such as that shown in Fig. 5.1.

New Concepts

Name	Comment
inorder traversal	similar to preorder traversal
Newick tree format	standard tree notation
postorder traversal	similar to inorder traversal
preorder traversal	method of visiting each node of a tree once
recursive structure	trees are recursive structures

New Programs

Name	Source	Help
`genTree`	book website	`genTree -h`
`new2view`	book website	`new2view -h`
`traverseTree`	book website	`traverseTree -h`

Problem 370 Begin by making the directory `TreesOfLife` and change into it. Here is a simple phylogeny:

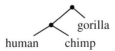

© Springer International Publishing AG 2017
B. Haubold and A. Börsch-Haubold, *Bioinformatics
for Evolutionary Biologists*, https://doi.org/10.1007/978-3-319-67395-0_5

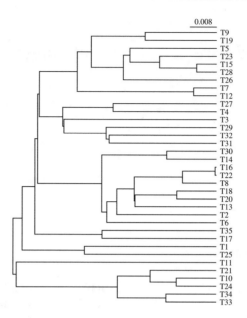

Fig. 5.1 A random tree of 35 taxa

Its textual representation is

```
((human:1,chimp:1):1,gorilla:1);
```

This notation is called the Newick format after "Newick's Lobster House" in New Hampshire, where the American Biologist Joe Felsenstein and a few colleagues adopted it on June 26, 1986. Translate the following tree to Newick and save it in `first.tree`.

A   B   C   D

Problem 371 The advantage of a standard tree notation is that programmers can write visualization software based on it. Use `new2view` to visualize `first.tree`.

Problem 372 The Newick format does not require branch lengths. Write `first.tree` without branch lengths, and save it as `second.tree`. How does `new2view` render branches without lengths?

Problem 373 In a Newick tree, any node can be either labeled or unlabeled. Remove the leaf labels from `second.tree` and label its root as "root". Save the result in `third.tree`.

Problem 374 Return to the tree with labeled leaves in `second.tree`. What happens when the labels within the pairs A, B, and C, D are switched? Does this change the biological interpretation of the tree? Save this tree as `fourth.tree`.

Problem 375 Use `new2view` to look at unrooted versions of all four trees generated so far. What is the difference to the rooted layout?

Problem 376 In biology, we often talk about "unrooted" trees, even though this is an oxymoron for computer scientists and mathematicians. To get a clearer understanding of what biologists mean when they refer to unrooted trees, save the code of the four trees in `fourTreesU.tree`. Then unroot them manually by editing each one such that their "root" has three children instead of the usual two. Use `new2view` to look at the unrooted trees in default layout and in rooted layout.

Problem 377 The bracket notation emphasizes that trees consist of subtrees, for example,

`(((,),),(,));`

It is a bit difficult to imagine what this tree looks like, but we can again use `new2view`, and this time label all nodes with their internal identifier by using the `-1` option

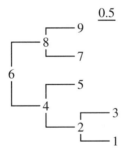

The subtree rooted on node 6 consists of subtrees rooted on nodes 8 and 4, and so on. This "recursive" structure leads to a method for traversing a tree, where the function `traverse` calls itself:

```
traverse(node)
    visit(node)
    traverse(leftChild(node))
    traverse(rightChild(node))
```

When we apply this procedure to node 6 in our tree, the nodes are visited in the order: 6, 4, 2, 1, 3, 5, 8, 7, and 9. Since the root is always visited first, then each one of the child nodes in turn, the procedure is called *preorder traversal*. In what order are the nodes of Fig. 5.2 visited during preorder traversal?

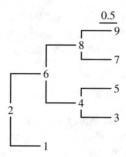

Fig. 5.2 A tree with all nodes labeled

Problem 378 Instead of preorder, a tree can also be traversed inorder by first visiting the left child, then the root in the middle, and finally the right child:

```
traverse(node)
    traverse(leftChild(node))
    visit(node)
    traverse(rightChild(node))
```

In what order does this algorithm visit nodes when applied to the root of Fig. 5.2?

Problem 379 Finally, the root can be visited last, which is called postorder traversal:

```
traverse(node)
    traverse(leftChild(node))
    traverse(rightChild(node))
    visit(node)
```

In what order does this algorithm visit nodes when applied to the root of Fig. 5.2?

Problem 380 Instead of traversing the tree in Fig. 5.2 by hand, write its topology without any labels into the file `traverse.tree` and use `traverseTree` to traverse it in the three possible modes. Can you see why they are also called *depth first* traversals?

Problem 381 Sometimes it is useful to quickly generate a tree, for example to test software that draws trees. The program `genTree` generates random phylogenies. By default, the branches of these trees have lengths proportional to the number of mutations along the branches. Run `genTree` with default options and visualize its output. The branches in the tree do not all end at the same vertical line, which indicates the present. Can you think of a reason for this heterogeneity in the positions of the leaves?

Problem 382 Generate two random phylogenies without mutations. How do they differ?

Problem 383 There are $n \times (n-1) \times \ldots \times 2 = n!$, that is, n-factorial, ways of ordering n objects. Write an AWK program for computing $n!$ as a function of $n = 1, 2, \ldots, 100$ and plot the result to get an idea of the number of possible phylogenies.

Problem 384 There are many trees that correspond to the same ordering of leaves. The number of phylogenies with n leaves is, therefore, probably larger than $n!$. Exactly how large, can be computed based on the following consideration [17, p. 20ff]: Two taxa are connected by a single rooted tree:

The third taxon, C can be added to any of the three edges e_1, e_2, and e_3, giving three trees:

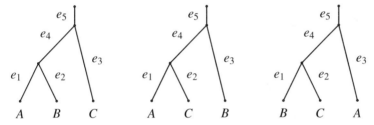

The fourth taxon can be added to any one of the five edges $e_1, e_2, \ldots, e_5$, yielding $3 \times 5 = 15$ rooted trees. In general, the number of rooted, bifurcating trees for n taxa is

$$3 \times 5 \times \ldots \times (2n - 3).$$

Write the program `numTrees.awk` to compute the number of rooted phylogenies as a function of n. Plot the result for $n = 2, 3, \ldots, 100$ and compare it to $n!$.

Problem 385 Generate another random tree with no mutations, but this time display branch lengths that are proportional to the speciation time.

Problem 386 When reconstructing phylogenies, biologists often talk about a *molecular clock*. Under the molecular clock model, mutations occur with constant rate along all branches of a phylogeny. Trees generated by `genTree` have this property. But in contrast to the clocks of everyday life, the molecular clock is stochastic. To emphasize the difference, compare two random trees with default mutations, one with branch lengths proportional to mutations (molecular clock), the other with branch lengths proportional to time (usual clock). Use the same seed for the random number generator to make the trees comparable.

5.2 Rooted Phylogeny

How can we calculate phylogenies from data rather than just simulate them? To construct a rooted phylogeny, start from a distance matrix like that shown in the top panel of Fig. 5.3a. Cluster the most similar organisms, A and B, such that the distance between A and its parent is half that between A and B (Fig. 5.3a, bottom panel). Then recompute the distance matrix with the new cluster, (A, B). Distances between groups of taxa are the average of the distances between the members of the groups. For example, the distance between C and (A, B) is $(4 + 4)/2 = 4$. This averaging of distances gives the method its name: "Unweighted Pair-Group Method using an Arithmetic average," or UPGMA [17, 46]. The most similar pair of taxa is chosen from the new matrix and clustered again (Fig. 5.3b). The cycle of matrix adjustment and clustering is repeated one more time to yield the final rooted phylogeny (Fig. 5.3c).

New Concept

Name	Comment
Unweighted Pair-Group Method using an Arithmetic average	clustering method

New Programs

Name	Source	Help
andi	book website	-h
clustDist	book website	clustDist -h
dnaDist	book website	dnaDist -h
gd	book website	gd -h
lscpu	system	man lscpu

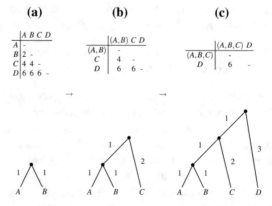

Fig. 5.3 Construction of a rooted phylogeny in three steps (**a–c**) using the "Unweighted Pair-Group Method using an Arithmetic average," UPGMA

Problem 387 Use UPGMA, paper, and pencil to draw the phylogeny implied by the distance matrix

	A	B	C	D
A	-			
B	6	-		
C	2	6	-	
D	6	4	6	-

Problem 388 There might be discrepancies between a UPGMA tree and the distances from which it was constructed. Perfect agreement is also known as "ultrametricity." Ultrametric distances fulfill the criterion that for three distances between taxa A, B, C we can find a labeling such that $d_{A,B} \leq d_{A,C} = d_{B,C}$. This "three point criterion" ensures that all leaves fall on a horizontal line, the present:

An ultrametric tree implies not only a molecular clock, but also that this stochastic clock behaves like a conventional clock. These strong assumptions underlie the simplicity of UPGMA. Are the distances in Problem 387 ultrametric?

Problem 389 Let's construct a phylogeny from some real sequence data. The file hominidae.fasta contains the extant genera of the *Hominidae*, also known as "great apes".

Problem 390 Look up the trivial names of these organisms in the taxonomy database on the NCBI website at

http://www.ncbi.nlm.nih.gov/taxonomy

Problem 391 We now recapitulate in detail how to get from sequence data to a tree. The sequences in hominidae.fasta are already aligned. Use gd, getSeq, and cutSeq to extract the first ten polymorphic positions in the alignment. From this, count by hand the pairwise differences between these taxa.

Problem 392 Use UPGMA, pencil, and paper to construct the phylogeny from the distances. Write the final tree in Newick format. Then enter the distances in the file test.dist using the format

```
4
name1 d_11 d_12 d_13 d_14
name2 d_21 d_22 d_23 d_24
name3 d_31 d_32 d_33 d_34
name4 d_41 d_42 d_43 d_44
```

Apply clustDist to check the intermediate matrices and final tree. Visualize the tree with new2view.

Problem 393 Use the program dnaDist to compute the pairwise distances from hominidae.fasta. Are the distances ultrametric?

Problem 394 Use clustDist with UPGMA to compute the *Hominidae* phylogeny from hominidae.dist. Pipe the result through new2view to visualize the tree.

Problem 395 Next, we compute the phylogeny of primates from their mitochondrial genome sequences. How many sequences are contained in primates.fasta and what is the range of their lengths?

Problem 396 The program andi [23] can quickly compute distances between genomes. Use andi to compute the pairwise distances between the primate mitochondria. Andi is a "multi-threaded" program, which means it can use more than one CPU at a time to carry out its computation. To find out the number of threads available, enter

```
lscpu
```

and look for the line starting

```
CPU(s):
```

Say, there are eight CPUs, then run the mitochondria computation with 1, 2, 4, and 8 threads and use the program time to measure the run time; for two threads this would be

```
time andi -t 2 primates.fasta
```

Problem 397 Cluster the distances using UPGMA and visualize the primate phylogeny with new2view. Do you get the same branching order for human, chimp, gorilla, and orangutan as with the *Hominidae* data set used in Problem 392?

5.3   Unrooted Phylogeny

All groups of organisms have a common ancestor, which means that all real phylogenies are rooted. The UPGMA algorithm we used in Sect. 5.2 is popular because it generates rooted phylogenies. The underlying assumption that the mutation rate ticks like a real clock along all branches fits the intuition that the leaves of a phylogeny should all be located in the present time. However, the mutation rates might vary along branches as shown in Fig. 5.4a. If we apply UPGMA to the distance matrix in Fig. 5.4b, *B* and *C* are clustered first, giving us the wrong tree. In this section, we look at the neighbor-joining algorithm [42], which overcomes this limitation of UPGMA and recovers the correct tree at the cost of losing the root node (Fig. 5.4c). However, we can reintroduce the root by placing it in the middle between the two most divergent taxa [16], a procedure known as "midpoint rooting".

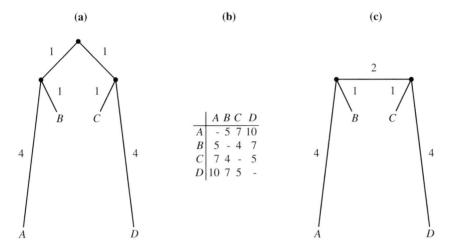

Fig. 5.4 Tree with varying mutation rates (**a**) and the corresponding distance matrix (**b**). By removing the root node from (**a**) we get (**c**)

New Concepts

Name	Comment
midpoint rooting	place root in the middle between most distant leaves
neighbor joining method	clustering method for distance data

New Program

Name	Comment	Help
retree	book website	menu-driven

Problem 398 Trees with varying mutation rates along their branches like in Fig. 5.4a result in distances that are called additive. Of the six distances among organisms $A, B, C,$ and D, four cross the root of the tree, $d_{AC}, d_{AD}, d_{BC},$ and d_{BD}. These can be divided in two groups of two, the sums of which are identical, $d_{AC} + d_{BD} = d_{AD} + d_{BC}$. These sums are always larger than the sum of the remaining two distances, d_{AB} and d_{CD}. This yields the "four point criterion" for additive distances:

$$d_{AC} + d_{BD} = d_{AD} + d_{BC} \geq d_{AB} + d_{CD}.$$

Check manually that the distances in Fig. 5.4b are additive.

Problem 399 List three taxa from Fig. 5.4a whose distances contradict the three point criterion of ultrametricity. This posits that given a tree, any triplet of leaves can be labeled such that

$$d_{AB} \leq d_{AC} = d_{BC}.$$

Problem 400 To trace the neighbor-joining algorithm, note down the top triangle of the distance matrix in Fig. 5.4b:

	A	B	C	D	r_i
A	-	5	7	10	
B		-	4	7	
C			-	5	
D				-	

The last column, r_i, is reserved for the row sums. Compute by hand the values of r_i, remembering that the distance matrix is symmetrical.

Problem 401 In the next step of neighbor-joining we compute for each pair of taxa, i, j, the difference between its distance, d_{ij}, and the normalized sum of the corresponding row sums, r_i, r_j:

$$S_{ij} = d_{ij} - \frac{r_i + r_j}{n - 2},$$

where n is the number of taxa. Compute the S_{ij}-values by hand and write them in the lower triangle of the distance matrix.

Problem 402 Instead of clustering the pair of taxa with the smallest distance as in UPGMA, neighbor-joining clusters the taxa with the smallest S_{ij}. If the new cluster is called c, the distances between c and some other cluster, k, is

$$d_{kc} = (d_{ik} + d_{jk} - d_{ij})/2.$$

The other distances are unchanged. Cluster one of the two pairs with the smallest S_{ij} and adjust the distance matrix accordingly.

Problem 403 We still need to know the lengths of the branches connecting the new cluster c and leaves i and j:

$$d_{ic} = \frac{(n - 2)d_{ij} + r_i - r_j}{2(n - 2)},$$

and

$$d_{jc} = \frac{(n - 2)d_{ij} + r_j - r_i}{2(n - 2)}.$$

Compute (by hand) the branch lengths for the new cluster.

Problem 404 To summarize, the neighbor-joining algorithm consists of four steps; given distances d_{ij},

• compute

$$r_i = \sum_j d_{ij},$$

- compute

$$S_{ij} = d_{ij} - (r_i + r_j)/(n - 2),$$

- cluster pair of taxa with smallest S_{ij} in node c with

$$d_{kc} = (d_{ik} + d_{jk} - d_{ij})/2,$$

- calculate branch lengths

$$d_{ic} = \frac{(n - 2)d_{ij} + r_i - r_j}{2(n - 2)}$$

$$d_{jc} = \frac{(n - 2)d_{ij} + r_j - r_i}{2(n - 2)}$$

This procedure is repeated until there are only three clusters left, i, j, k, which is the stage we have reached in our example. These are joined to the pseudo-root r, by branches with the following lengths

$$d_{ri} = (d_{ij} + d_{ik} - d_{jk})/2$$
$$d_{rj} = (d_{ji} + d_{jk} - d_{ik})/2$$
$$d_{rk} = (d_{ki} + d_{kj} - d_{ij})/2$$

What are the lengths of the last three branches added to our example tree?

Problem 405 Create the directory NeighborJoining for this session and change into it. Enter our example distances (Fig. 5.4) in file test.dist and use clustDist to automatically trace the steps just completed by hand. Save the tree as testU.tree and draw it using new2view.

Problem 406 Root the example tree by placing the root in the middle of the longest path from one leaf to another. Write down the midpoint-rooted phylogeny in Newick format and save it to testR.tree. Then draw it with new2view.

Problem 407 Use the PHYLIP program retree to midpoint-root our unrooted example tree from Problem 405, which was saved as testU.tree. Save the final tree in rooted form. Does retree return the same result as the manual midpoint rooting?

Problem 408 Use andi together with clustDist to compute the neighbor joining tree of the mitochondrial genomes in primates.fasta. Midpoint-root this version of the primate phylogeny and compare it to the UPGMA tree (Problem 396). Can you spot any differences?

Chapter 6
Evolution Within Populations

6.1 Descent from One or Two Parents

Every one of us has two parents, four grandparents, eight great-grandparents, and so on. Figure 6.1 shows a random ancestry for a single individual in a population of seven individuals. The individuals are diploid, hence the two dots indicating two genes, and there are two sexes, ellipses, and boxes. As we shall see that the rapid growth of the number of ancestors has the curious effect that it is easy to have a famous forbear—but difficult to not share him with everybody [13, 41]. In contrast, each gene has a single ancestor, if we ignore recombination for now. Using simulations, we learn that in the uni-parental genealogies of genes it takes much longer to find common ancestors than in the bi-parental genealogies of individuals.

New Concepts

Name	Comment
bi-parental genealogy	the genealogy of people
uni-parental genealogy	the genealogy of genes
Wright–Fisher model	model of evolution within populations

New Programs

Name	Source	Help
drawGenealogy	book website	drawGenealogy -h
drawWrightFisher	book website	drawWrightFisher -h

6.1.1 Bi-Parental Genealogy

Problem 409 Write down the number of ancestors as a function of the number of generations back in time. How many ancestors do you have 30 generations back?

Problem 410 Trace the ancestry of individual i_4 in Fig. 6.1. Walk back in time from b_0 to b_3 and count the ancestors of i_4 in each generation. What do you observe?

© Springer International Publishing AG 2017
B. Haubold and A. Börsch-Haubold, *Bioinformatics for Evolutionary Biologists*, https://doi.org/10.1007/978-3-319-67395-0_6

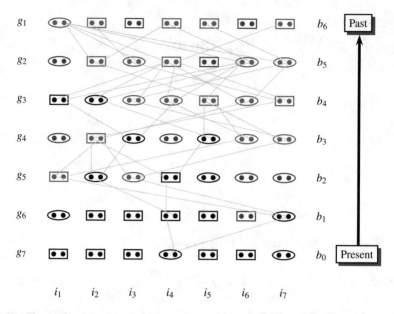

Fig. 6.1 Simulation of the ancestors of a single individual. Ellipses and boxes indicate two sexes, the dots two genes. g_i stands for generation i, b_j for j generations back

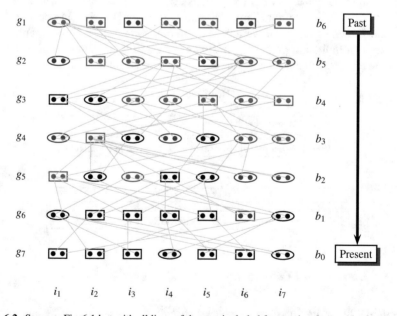

Fig. 6.2 Same as Fig. 6.1 but with all lines of descent included for moving forward in time

Problem 411 Figure 6.2 shows the same simulation as Fig. 6.1, but this time all lines of descent are included, not only those leading to i_4. So we can walk from any individual forward in time to visit all its descendants. By doing this, can you explain the color coding of black, blue, and red individuals?

Problem 412 Create the directory `Descent` and change into it. Figures 6.1 and 6.2 were drawn using the program `drawGenealogy`, which generates LATEX output. Use it to simulate new random family histories using options similar to the above Figures. Typeset and display the figures using the commands `latex`, `dvips`, and `gv`.

Problem 413 `drawGenealogy` also has a non-graphical mode. Use it to determine the number of generations until the first universal ancestor appears for populations of sizes 10, 20, 50, 100, 200, 500, and 1000. Compute averages over 100 iterations every time and plot the result. Compare your results to the expectation of $\log_2(N)$, where N is the population size [13]. This can be specified in `gnuplot` by

```
f(x) = log(x) / log(2)
plot f(x)
```

Problem 414 Use `drawGenealogy` to determine the average number of generations until all present-day individuals have identical ancestors. Compare your results to the expectation of $1.77 \times \log_2(N)$ [13].

6.1.2 Uni-Parental Genealogy

Problem 415 In contrast to individuals, who have two parents, genes only have one. Figure 6.3 shows the third version of Fig. 6.1, where the lines of descent are restricted to those of the two genes in individual i_4. What can you say about the common ancestor of those two genes? How does this compare to the ancestry of individuals?

Problem 416 Populations are often modeled as consisting of genes without reference to individuals or gender. This abstract model of a population is called the Wright–Fisher model in honor of two founding figures of population genetics, Ronald A. Fisher (1890–1962) and Sewall Wright (1889–1988). Figure 6.4 shows two generations in the evolution of a Wright–Fisher population. To get from one generation to the next, ancestors are simply picked at random, as indicated by the arrows. To see how this works, extend Fig. 6.4 for one generation by manually drawing another set of eight genes in g_3. To determine the ancestor of the first gene, use

```
awk -v seed = $RANDOM 'BEGIN {srand(seed); print int(
      rand() * 8 + 1)}'
```

and draw the appropriate arrow. Then repeat for the remaining genes.

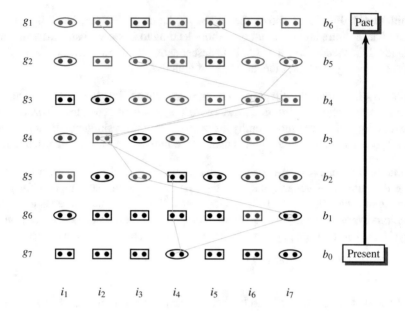

Fig. 6.3 Tracing the genes of individual i_4

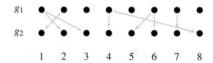

Fig. 6.4 Two generations of a population size 8 under the Wright–Fisher model

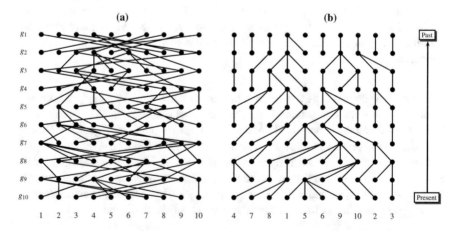

Fig. 6.5 Tangled (**a**) and untangled (**b**) versions of the same Wright–Fisher simulation

Fig. 6.6 Two generations of a population under the Wright–Fisher model consisting of eight genes

Problem 417 Figure 6.5a depicts ten generations of a Wright–Fisher population consisting of ten genes. With all the crisscrossing ancestral lines connecting the genes between generations, it is a bit hard to see what is going on. It is possible to untangle the lines of descent such that no two lines cross. Figure 6.5b shows the untangled version of Fig. 6.5a. This makes it much simpler to trace ancestry back in time. Does Fig. 6.5 contain a common ancestor of all genes?

Problem 418 Use the program `drawWrightFisher` to simulate Wright–Fisher populations in LATEX. With a population size of ten, how often do you observe a common ancestor for the entire population within ten generations back? Use `-a` for ancestor tracing, `-t` for wrapping the LATEX code, and carry out, say, ten trials.

Problem 419 When going one generation back in time, the number of lineages that eventually end up in the common ancestor either stays the same, or decreases. For example, in Fig. 6.6 that number goes from originally eight in b_0 to five in b_1. You can think of each gene (dot) in Fig. 6.6 as randomly picking its ancestor in the preceding generation. Occasionally two genes pick the same ancestor. What is the probability of this occurring?

Problem 420 If two genes pick an identical ancestor, their lineages fuse and the number of ancestral lineages is reduced by one. To simulate such "ancestor events", we first generate N random integers between 0 and $N - 1$:

```
BEGIN{
    # seed the random number generator
    srand(seed)
    for(i=0; i<N; i++)
        print int(rand()*N)
}
```

Copy this code in the file `trace1.awk`, and when running it, pass the value of N, and a unique seed for the random number generator:

```
awk -f trace1.awk -v seed=$RANDOM -v N=10
```

Use `sort`, `unique`, and `wc` to count the number of ancestral lineages remaining after going back one generation. Repeat this a couple of times. Do you ever observe ten lineages remaining?

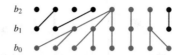

b_2
b_1
b_0

Fig. 6.7 The Wright–Fisher population of Fig. 6.6 traced back one more generation

Problem 421 We saw in Problem 419 that the probability of two genes having the same ancestor in the previous generation is $1/N$. So the probability of two genes *not* having a common ancestor is $1 - 1/N$. The probability of three genes not having the same ancestor is

$$\left(1 - \frac{1}{N}\right)\left(1 - \frac{2}{N}\right)$$

and hence the probability of N genes not having the same ancestor is

$$P_{\mathrm{n}} = \left(1 - \frac{1}{N}\right)\left(1 - \frac{2}{N}\right)\dots\left(1 - \frac{N-1}{N}\right). \tag{6.1}$$

What is P_{n} for $N = 10$? Check your result through simulation.

Problem 422 Figure 6.7 shows two steps back in time rather than just the single one in Fig. 6.6. If we wish to simulate this second step, we need to distinguish the population size from the number of lineages that end up in the present, or ancestral lineages. We have already seen that after the first generation the number of ancestral lineages is usually smaller than the population size. We now wish to simulate the effect of going back one step when starting from an arbitrary number of ancestral lineages, n, given the population size, N. Here is an edited version of `trace1.awk` for doing this:

```
BEGIN{
    srand(seed)
    # initialize the population
    for(i=0; i<N; i++)
        pop[i] = 0
    # sample with replacement
    for(i=0; i<n; i++){
        j = int(rand()*N)
        pop[j] = 1
    }
    # count the number of lineages
    nn = 0;
    for(i=0; i<N; i++)
        nn += pop[i]
    print nn
}
```

Save your version as `trace2.awk` and run it a few times. How often do you observe a reduction in n if you start with $N = 100$ and $n = 10$?

Problem 423 To calculate the probability of an ancestor event when considering a small number of n lineages in a large population of N genes, we can modify Eq. 6.1 to get

$$P_n = \left(1 - \frac{1}{N}\right)\left(1 - \frac{2}{N}\right)\dots\left(1 - \frac{n-1}{N}\right).$$

Since N is large, we can ignore terms with N^{-2} or smaller to get

$$P_n \approx 1 - \frac{1}{N} - \frac{2}{N} - \dots - \frac{n-1}{N}.$$

The complement of this is the probability of an ancestor event

$$P_a = \frac{1 + 2 + \dots + n - 1}{N} = \frac{n(n-1)}{2N}.$$

What is the probability of observing an ancestor event if $N = 1000$ and $n = 10$? Check your result through simulation.

Problem 424 Finally, we would like to trace the lineages back to their common ancestor. As we have just seen, once the number of ancestral lineages is much smaller than the population size, any further reduction in lineages takes many generations on average. To simulate exactly how many, wrap the code in `trace2.awk` in a `while` loop that repeats until the number of lineages is 1; in other words, until the most recent common ancestor of n genes has been reached.

```
BEGIN{
    srand(seed)
    g = 0    # number of generations
    while(n > 1){
        g++
        for(i=0; i<N; i++)
            pop[i] = 0
        for(i=0; i<n; i++){
            j = int(rand()*N)
            pop[j] = 1
        }
        nn = 0   # new n
        for(i=0; i<N; i++)
            nn += pop[i]
        if(nn < n)
            print g "\t" nn
        n = nn
    }
}
```

Save this as `trace3.awk` and plot the number of lineages as a function of the number of generations for $N = 100$ and $n = 100$.

Problem 425   Run `trace3.awk` a couple of times. What happens to the time to the most recent common ancestor as you switch between the two most extreme values for n possible, $n = 2$ and $n = N$?

Problem 426   The expected time to the most recent common ancestor is [49, p. 76]

$$E\{T_{\text{MRCA}}\} = 2N \left(1 - \frac{1}{n} \right). \tag{6.2}$$

What is the expected TMRCA for $n = 2$ and $n \to \infty$?

Problem 427   Simulate the average T_{MRCA} from 100 replicates for $n = 2, 5, 10, 20$ with $N = 100$ by writing the script `simTmrca.sh` to drive `trace3.awk`. Compare your results to the theoretical expectation in Eq. 6.2.

6.2   The Coalescent

Figure 6.8a shows a population of ten genes evolving for ten generations under the Wright–Fisher model. Investigations of real genes are usually restricted to small samples, say the three genes marked in blue. By untangling the lines of descent, we can see in Fig. 6.8b that these three genes are connected by a tree. If we just concentrate on this tree, we can further reduce it to the nodes where two lines of descent collide as we move from the present into the past. A different way of looking at such a collision is to say that two lines of descent merge or "coalesce" into one.

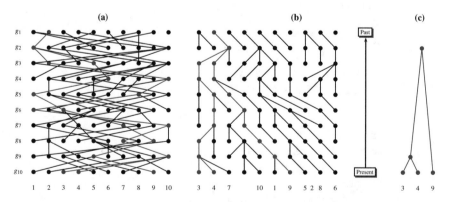

Fig. 6.8   A population under the Wright–Fisher (**a**), its untangled version (**b**), and the coalescent for three of its lineages (**c**)

The collection of such coalescence events is depicted in Fig. 6.8c and is called the
"coalescent". It describes the descent of a sample of genes evolving under the Wright–
Fisher model from the preset to the most recent common ancestor. As we shall see,
the coalescent is a tool for simulating the evolution of genes.

New Concepts

Name	Comment
coalescent	genealogy of gene sample under Wright–Fisher model
Watterson's equation	mutations as a function of population and sample size

New Programs

Name	Source	Help
coalescent.awk	book website	coalescent.awk -v h=1
ms	book website	ms
rpois.awk	book website	rpois.awk -v h=1
watterson	book website	watterson -h
tabix	book website	man tabix

Problem 428 Coalescents are random trees or genealogies. To construct a coalescent, we represent it as an array

Index	1 2 3 4 5 6 7
Node	1 2 3 4 5 6 7

where nodes 1–4 refer to leaves (green) and nodes 5–7 to inner nodes (black). Draw
an example tree with these nodes. What is the sample size, n, for this coalescent?

Problem 429 All nodes of a coalescent are annotated with times. The leaves get
times 0:

Index	1 2 3 4 5 6 7
Node	1 2 3 4 5 6 7
Time	0 0 0 0

To compute the times of the inner nodes, we divide the coalescent into intervals T_i,
where i is the number of lines of descent in the tree during that time (Fig. 6.9). Where
would T_1 be in this graph, and why is it not shown?

Problem 430 To compute T_i, start from the probability of a coalescence event
derived in Probelm 423:

$$P_{\mathrm{a}} = \frac{n(n-1)}{2N}.$$

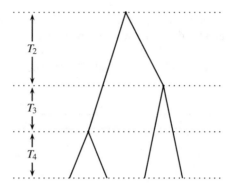

Fig. 6.9 Coalescent with time intervals T_i. These denote the regions of the tree consisting of i lines of descent

This means the time to the next coalescent event is an exponentially distributed random variable with mean $1/P_a$. Since we only know the sample size n and not the population size N, we measure time in units of $2N$ generations and compute T_4

```
awk -v s = $RANDOM 'BEGIN {srand(s); i = 4; r = -2*
   log(1-rand())/i/(i-1);print r}'
```

Use this code to compute T_i for $i = 4, 3, 2$, and fill in the coalescence times in the above table. What is the time to the most recent common ancestor in your coalescent?

Problem 431 We can automate the computation of coalescence times

```
BEGIN{
     srand(seed)
     t=0
     for(i=n;i>=2;i--){
          t -= 2*log(1-rand())/i/(i-1)
          print i, t
     }
}
```

Copy this code into the file genCoalTimes.awk and then run it

```
awk -v seed=$RANDOM -v n=10 -f genCoalTimes.awk
```

Wrap this code in a script to determine the average time to the most recent common ancestor (T_{MRCA}) from 100 iterations with $n = 2$ and $n = 1000$.

Problem 432 Having learned how to assign times to nodes, we next learn how to construct a random tree from them—the coalescent. This requires shuffling a set of nodes. Consider the five nodes 1, 2, 3, 4, 5; after shuffling, they might have the order 4, 5, 1, 2, 3. The way to achieve this for n nodes in node array a is as follows:

1. Pick a random number r between 1 and n.

2. Swap the $a[r]$ and $a[n]$.
3. Reduce n by 1.
4. Repeat.

This technique is also known as sampling without replacement. It depends on distinguishing between the value of an index, i, and the value of an array, a, at that position, $a[i]$. Shuffle 1, 2, 3, 4, 5 using the random indexes 1, 3, 1, 2.

Problem 433 To construct a coalescent, we need to apply the shuffling procedure to the array of leaves and internal nodes. For this, we add three auxiliary rows to our table so we can overwrite node labels:

Index	1 2 3 4 5 6 7
Node	1 2 3 4 5 6 7
Child1	
Child2	
Time	0 0 0 0

We begin with the first internal node, 5, and pick a random child among the four leaves:

```
awk -v s = $RANDOM 'BEGIN{srand(s); print int
    (rand()*4+1)}'
```

Say, this is 1; so the fist child of 5 is 1 and we replace node 1 by the leftmost leaf in this round, 4:

Index	1 2 3 4 5 6 7
Node	1 2 3 4 5 6 7
	4
Child1	1
Child2	
Time	0 0 0 0

We draw another random number, but this time only the first three leaves are candidates

```
awk -v s = $RANDOM 'BEGIN{srand(s); print int
    (rand()*3+1)}'
```

Say that is 2; then the second child of 5 is 2. In addition, 2 is replaced by 5:

Index	1 2 3 4 5 6 7
Node	1 2 3 4 5 6 7
	4 5
Child1	1
Child2	2
Time	0 0 0 0

So in each round the first child of node v is replaced by the leftmost leaf available in that round, and the second child is replaced by v. Algorithm 3 summarizes these steps. Finish the tree construction.

Algorithm 3 Generate coalescent

Require: n {sample size}
Require: tree {array of n leaves followed by $n - 1$ internal nodes}
Ensure: Tree topology
1: **for** $i \leftarrow n$ **to** 2 **do**
2:     $p \leftarrow i \times \text{ran}() + 1$ $\{1 \leq p \leq i\}$
3:     $\text{tree}[2 \times n - i].\text{child1} \leftarrow \text{tree}[p]$
4:     $\text{tree}[p] \leftarrow \text{tree}[i - 1]$ {Replace child by the rightmost entry in the "leafy" part of tree}
5:     $p \leftarrow (i - 1) \times \text{ran}() + 1$ $\{1 \leq p \leq i - 1\}$
6:     $\text{tree}[2 \times n - i].\text{child2} \leftarrow \text{tree}[p]$
7:     $\text{tree}[p] \leftarrow \text{tree}[2 \times n - i]$ {Replace child by parent}
8: **end for**

Problem 434 Sketch the coalescent just constructed.

Problem 435 Picking children can be automated:

```
BEGIN{
    srand(seed)
    print "# Pa\tC1\tC2"
    for(i=n; i>=2; i--){
        child1 = int(rand()*i+1)
        child2 = int(rand()*(i-1)+1)
        print 2*n-i+1 "\t" child1 "\t" child2
    }
}
```

Save this code in the program pickChildren.awk and use it together with genCoalTimes.awk to generate a coalescent for $n = 4$.

Problem 436 At this stage, we have coalescence times and a branching order. To achieve biological relevance, we still need mutations. They are generated as Poisson-distributed random variables for each branch with expectation

$$\lambda = t\theta/2,$$

where t is the branch length and $\theta = 4N\mu$, where μ is the number of mutations per generation per site. Use the program rpois.awk to generate the random variable, for example

```
awk -f rpois.awk -v lambda=0.8
```

As our coalescent has six branches, six mutation counts need to be generated. Compare the total number of mutations on the coalescent with the corresponding expectation according to Watterson's equation [50]

$$E\{S\} = \theta \sum_{i=1}^{n-1} \frac{1}{i},$$

where $E\{S\}$ is the expected number of segregating sites. This equation is implemented in the program watterson.

Problem 437 Use the program coalescent.awk to simulate two haplotype samples of size 4 with $\theta = 10$. Can you interpret the output?

Problem 438 The program coalescent.awk can also print the coalescent tree underlying the simulation of SNPs. Run it for single sample, extract the coalescent with grep, and visualize it with new2view. Repeat a few times to get a visual impression of the coalescent. What is the meaning of the scale?

Problem 439 The program ms is often used for real coalescent simulations [25]. Use it to generate two samples of size 4 with $\theta = 10$, for example,

```
ms 4 2 -t 10

ms 4 2 -t 10
60977 30522 51696

//
segsites: 6
positions: 0.0675 0.1627 0.2495 0.2952 0.3512 0.4482
010111
101000
010111
010111

//
segsites: 12
positions: 0.1269 0.2146 0.3704...
000001000010
000001000010
111110111101
000000000000
```

Row by row this output means:

- Row 1: Repetition of the command
- Row 2: Initialization of the random number generator
- Row 4: Start of the first sample
- Row 5: Number of segregating sites (mutations): 6
- Row 6: Positions of the mutations along the interval (0, 1)
- Rows 7–10: Haplotypes; 0 indicates ancestral state, 1 mutant

Use ms to simulate 10^4 samples of size 4 with $\theta = 10$. Use grep and AWK to estimate the average number of mutations. Compare this to the expectation value.

Problem 440 Let us say we observe a single mutation in a sample of four aligned DNA sequences, far fewer mutations than the number expected with $\theta = 10$. How significant is the deviation between observation and expectation?

Mouse Genome

Problem 441 Next, we investigate single nucleotide polymorphism (SNP) data in mice. We concentrate on chromosome 19, the smallest mouse chromosome, which was sampled from 17 mice. To query the SNPs in the first 5 Mb of chromosome 19, enter

```
tabix http://guanine.evolbio.mpg.de/problemsBook/chr19.
    mgp.vcf.gz 19:1-5000000
```

What is the location of the first and the last SNP on chromosome 19?

Problem 442 Count the SNPs located between the first and the last SNP (wc -l), and infer from this the number of SNPs per nucleotide. What is θ per nucleotide? (Remember, 17 mice were sequenced.)

Problem 443 Count the SNPs in mice between position 5,189,001 and 5,190,000. How many SNPs are expected for this interval? Is the deviation between theory and observation significant?

Chapter 7
Additional Topics

7.1 Statistics

Compared to other branches of mathematics, statistics is a young discipline. Take Student's t-test, which assesses the hypothesis that two small samples are drawn from the same population. It was published just a century ago by William S. Gosset (1876–1937), who used "Student" as a pseudonym for his work in statistics [45]. Gosset trained as a chemist and worked all his life in the management of the Guinness brewery, first in Dublin and later in London. A central aim of the company leadership at the time was to make brewing "scientific". This required above all experimentation on such things as the effect of the resin content of hops on beer quality. However, once the relevant measurements had been made, a structured approach to their interpretation was also needed; how large a difference in hop resin content made a significant difference to the lifetime of stout? Today we call the investigation of such questions "statistics" and Gosset was one of its pioneers. Rather than the resin content of hops, we take as our example an investigation of the effect of fatty food on gene expression in mice [10].

New Concepts

Name	Comment
false discovery rate	false-positive rate among rejected hypotheses
gene ontology	functional categories for all genes
mouse transcriptome	all transcripts of the mouse genome
type I error (false-positive rate)	reject true null hypothesis
type II error (false-negative rate)	don't reject false null hypothesis

New Programs

Name	Source	Help
simNorm	book website	simNorm -h
testMeans	book website	testMeans -h

© Springer International Publishing AG 2017   127
B. Haubold and A. Börsch-Haubold, *Bioinformatics*
for Evolutionary Biologists, https://doi.org/10.1007/978-3-319-67395-0_7

7.1.1   The Significance of Single Experiments

Problem 444 Eight mice were given standard food, and are called sample A. Eight mice were given fatty food, sample B. RNA was extracted from liver and quantified on hybridization chips. The files `all_a.txt` and `all_b.txt` contain the results for experiments A and B, respectively. We start by investigating a single gene, *Plin5*. Use `grep` to extract its expression levels and save them in files `plin5_a.txt` and `plin5_b.txt`.

Problem 445 Compute the average expression values of *Plin5* in both experiments.

Problem 446 Next we investigate the difference between the two averages using the program `testMeans` with default parameters. Is the difference between the estimated means significant?

Problem 447 The default method for calculating significance in `testMeans` uses a formula from Gosset's original work. However, this is based on the assumption that the two samples compared were drawn from a normal distribution. As this assumption may not hold, `testMeans` also provides a Monte Carlo test for computing P-values. Like gambling at the Monte Carlo Casino in Monaco, a Monte Carlo method in statistics is based on chance: Consider two samples, S_1 and S_2, and their means, m_1 and m_2. Then calculate the difference between the two means: $\Delta_0 = |m_1 - m_2|$. Now shuffle the elements of S_1 and S_2 between the sets and repeat the computation of their means and the difference between them. Repeat this n times to get $\Delta_1, \Delta_2, ...\Delta_n$. The significance of the difference between the two samples is the frequency with a $\Delta_i \geq \Delta_0, i = 1, 2, ..., n$. One implication of this method is that P cannot fall below $1/n$; in other words, the theoretical minimum of P depends on the number of shufflings we carry out. For this reason the user of `testMeans` can vary n. Compare the P-value obtained by `testMeans` using Monte Carlo and the P-value obtained using the default method.

Before comparing all the genes in experiments A and B, we investigate the statistics of multiple tests using simulated data.

7.1.2   The Significance of Multiple Experiments

Problem 448 The program `simNorm` generates samples drawn from a normal distribution. Use the program to generate 100 samples of size 8 with mean 12. Save the results in the file `experiment1.txt`. Repeat the simulation and save the results in `experiment2.txt`. Look at the first rows of the two files (`head`): They contain the values for sample 1, S1, followed by the values for sample 2, S2, and so on. We can interpret these samples as control/experiment for genes S1, S2, and so on. What is the number of false-positive results, the false-positive rate, if $\alpha = 0.05$ (`testMeans`)?

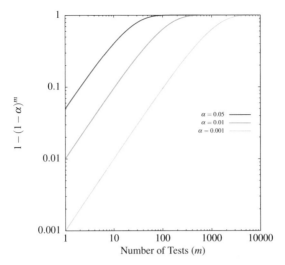

Fig. 7.1 A different view of the false-positive rate: The probability of obtaining at least one false-positive result as a function of the number of hypothesis tests

Problem 449 Repeat the estimation of the false-positive rate for 10^4 experiments.

Problem 450 As illustrated in Fig. 7.1, the more tests we carry out, the greater the probability that we obtain at least one false-positive. What is the probability of obtaining at least one false-positive with $\alpha = 0.05$ when carrying out 100 tests?

Problem 451 In statistics the false-positive rate is also known as the type I error. So far we set this to $\alpha = 0.05$ per experiment. However, we can also regard the 10^4 experiments as a single unit. Then we need to divide the original α by the number of tests carried out to assess the null hypothesis that the ensemble contains no sample with a significant difference. This division of α by the number of hypothesis tests is called the Bonferroni correction. What is the false-positive rate if we analyze our 10^4 experiments using the Bonferroni correction?

Problem 452 Now we simulate samples with different means. Run 10^4 experiments with $\mu = 6$ and $\sigma = 2.5$ and save them in `experiment1.txt`. Repeat the simulation with $\mu = 8$ and $\sigma = 2.5$, and save the result in `experiment2.txt`. What is the false-negative rate, β, if we leave $\alpha = 0.05$?

Problem 453 Repeat the simulation of samples with different means. Like before, run 10^4 experiments with $\mu = 6$ and $\sigma = 2.5$ and save them in `experiment1.txt`. For the second simulation, use $\mu = 12$ and $\sigma = 2.5$, and save the result in `experiment2.txt`. What is the false-negative rate this time?

Problem 454 Repeat the simulation of `exeriment2.txt` with $\mu = 12$ and the larger standard deviation $\sigma = 3.5$. How does β change?

Problem 455 Analyze `experiment1.txt` and `experiment2.txt` again, but this time use Bonferroni correction [40]. What is the false-negative rate, also known as type II error, now?

Problem 456 In genomics, we are often not primarily interested in eliminating the type I error, as this can lead to a large type II error. Hence the concept of false discovery rate, fdr, has been developed. The fdr is the fraction of false-positive results among the *rejected* hypotheses, rather than among all hypotheses. In order to set the fdr to some level δ, the original P-values are sorted $P_1 \leq P_2 \leq \ldots \leq P_m$; then P_j is significant if $P_j \leq \delta j / m$. This method is due to Benjamini–Hochberg [9] and hence also known as the Benjamini–Hochberg correction. Repeat the analysis using this correction. What is the type II error now?

Problem 457 Simulate two sets of 10^4 experiments with identical means $\mu_1 = \mu_2 = 6$ and standard deviations $\sigma_1 = \sigma_2 = 2.5$. Analyze the results using the Benjamini–Hochberg correction. How large is the type I error?

Problem 458 Our ability to detect an effect in an experiment depends on two quantities: effect size and sample size. To investigate sample size, simulate again 10^4 pairs of experiments with $\mu_1 = 6, \sigma_1 = 2.5$ and $\mu_2 = 12, \sigma_2 = 3.5$ for sample sizes of $n = 2, 5, 10, 20, 50$. Analyze the results using the Benjamini–Hochberg correction and plot the type II error, β, as a function of sample size, n.

Problem 459 Repeat the sample size simulation with a smaller effect size: $\mu_1 = 6, \sigma_1 = 2.5$, and $\mu_2 = 7, \sigma_2 = 2.5$. Since the effect size is small, carry out the simulation for a larger range of sample sizes: $n = 2, 5, 10, 20, 50, 100, 200, 500$. What sample size is necessary to drive β below 0.05? Hint: Plot a horizontal line in `gnuplot` using the syntax

```
f(x)=0.05
plot ..., f(x) t "" w l
```

7.1.3 Mouse Transcriptome Data

Problem 460 The files `all_a.txt` and `all_b.txt` contain the data for all the mouse transcriptome probes assayed in experiments A and B. How many probes were investigated (`wc -l`)? Some genes were assayed more than once. How many distinct genes were assayed (`cut, sort, uniq, wc -l`)?

Problem 461 Analyze the data and filter them using the Benjamini–Hochberg correction with the relatively permissive threshold of $\delta = 0.1$. Save the genes deemed significant in the file `genes.txt`. How many distinct genes does it contain?

Problem 462 To finish our analysis of mouse transcriptome data, we investigate whether the genes in in `genes.txt` are enriched for a particular function. Biological functions are codified *gene ontologies* (GO), which are hierarchical, for example genes involved in eye development are a subset of genes involved in development. Figure 7.2 shows the result of the GO analysis for our genes. Look for the highest node in the graph that is significant (red). Can you relate it to the underlying study?

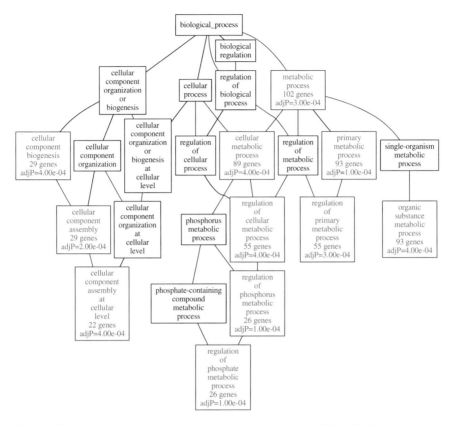

Fig. 7.2 Result of enrichment analysis using the gene ontology (GO) [47]. Computed using www.webgestalt.org

7.2 Relational Databases

In the late 1960s, the British mathematician Edgar F. Codd (1923–2003) proposed a new model for storing and accessing data. This model, called the relational data model, has become the standard way of dealing with large data sets. Originally used mainly in government and business, relational databases are now also ubiquitous in genomics. In this section we learn how to construct and query relational databases.

There are a number of software systems available for doing this and they all implement the query language SQL. However, there are differences, and the most important distinction is between systems with a client–server structure and those without. Systems with a client–server structure, such as Oracle, Mysql, and Postgresql, are usually centered on a server hosting one or more databases (Fig. 7.3). A potentially large number of clients connects via the internet to this server. As an example, we introduce and query the ENSEMBL database in this section. It contains genome data on vertebrates and is hosted on a public server under Mysql [5].

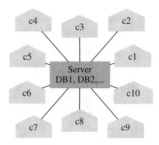

Fig. 7.3 A database server hosting several databases connected to ten clients

Server-client systems are powerful and correspondingly challenging to construct and administer, as opposed to mere querying. Fortunately, there are also simpler systems, where the database is just a local file. As an example for this kind of system we experiment with the Sqlite database.

New Concepts

Name	Comment
database client	program for accessing a database
database server	program for hosting a database
ENSEMBL	collection of vertebrate genome databases
Java	higher level programming language
relational databases	collections of data in tabular format
SQL	programming language for querying databases

New Programs

Name	Source	Help
javac/java	package manager	man java
mysql	package manager	man mysql
sqlite3	package manager	man sqlite3

7.2.1 Mouse Expression Data

Problem 463 The files `fatty_food.txt` and `normal_food.txt` contain a subset of the mouse transcriptome data we already used in Chap. 7.1. Figure 7.4 shows an Entity-relation (ER) model of the database we wish to construct. Boxes are *entities*, ellipses *attributes*, and the diamond denotes a *relationship*. It's a one-to-one relationship, where each entry in `fatty_food` has a corresponding entry in `normal_food`. The underlined attribute is a unique *primary key*. Here is the code for constructing table `normal_food`

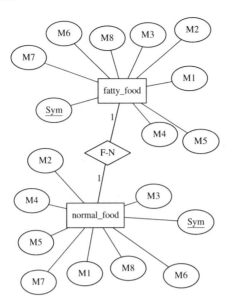

Fig. 7.4 Entity-relation (ER) model for `mouseExpressDb`

```
create table normal_food(
   Sym varchar(18),
   M1 float,
   M2 float,
   M3 float,
   M4 float,
   M5 float,
   M6 float,
   M7 float,
   M8 float,
   primary key(Sym)
);
```

Create the directory `RelationalDb`, and copy the data files into it. Write the code for constructing table `normal_food` into the the file `normal_food.sql`, and then write the corresponding file `fatty_food.sql`.

Problem 464 There is still one element of the ER-diagram in Fig. 7.4 missing from our SQL-code, the relationship. This is modeled by a *foreign key*, which is declared as

```
foreign key(x) references some_table(y)
```

where attribute x refers to attribute y in `some_table`. We wish to model the fact that for every entry in `fatty_food` there is also an entry in `normal_food`, but not necessarily vice versa. Add code for the corresponding foreign key to `fatty_food.sql`.

Problem 465 Next, we construct the actual database `mouseExpress.db` using `sqlite3`. The following commands should get you there:

- Log on to a database: `sqlite3 dbname`; if dbname does not yet exist, it is created
- Switch on foreign keys: `PRAGMA foreign_keys = ON;`
- Read in a file containing SQL commands: `.read <foo.sql>`
- Show existing tables: `.tables`
- We use tabs to delineate columns instead of the default `|`: `.separator "\t"`
- Import data contained in file `table.txt` into table `table`: `.import table.txt table`
- Show all attributes of the first ten entries in a table: `select * from table limit 10;`
- Quit `sqlite`: `.quit`

Problem 466 Instead of adding many rows to a table using `.import`, we can add individual rows using the `insert` command:

```
insert into normal_food
values ('toy_gene1', 1.0, 2.0, 3.0, 4.0, 5.0, 6.0,
    7.0, 8.0);
```

Notice the semi-colon that closes every SQL command. Unless an SQL command is closed, `sqlite` prints the prompt

```
...>
```

indicating that it awaits further input. This will also occur if, for example, the single quote surrounding `toy_gene` is not closed again. If you are stuck with `...>` and cannot get back to `sqlite>`, check to see what is keeping the command open. In most cases it will be a missing semi-colon. To retrieve the entry we just generated, enter

```
select * from normal_food where sym like 'toy_gene1';
```

Delete our toy entry

```
delete from normal_food where sym like 'toy_gene1';
```

and check the result

```
select * from normal_food where sym like 'toy_gene1';
```

Enter the following data to `normal_food`:

sym	m1	m2	m3	m4	m5	m6	m7	m8
toy_gene2	17.1	9.5	27.7	6.5	24.1	30.2	30.6	14.3

and delete them again.

Problem 467 What happens if we make a second entry for an existing `sym` like *Plin5*?

Problem 468 What happens if we try to enter '

sym	m1	m2	m3	m4	m5	m6	m7	m8
toy_gene3	3.4	8.0	4.4	26.7	8.6	26.6	4.8	20.5

into `fatty_food`?

7.2.2 SQL Queries

Problem 469 Recall form the database construction in Problem 465 that instead of entering commands interactively, they can also be read from files using `.read`. To experiment with this feature, write the four commands constructed in Problem 466 into the file `insert.sql` and enter them with `.read`. What happens if the `insert` command contains one value too many or too few?

Problem 470 Use the SQL-command `count` to determine the number of entries in `normal_food` and `fatty_food`.

Problem 471 We need one expression value per gene, the average of m1, m2, ..., m8. Construct a table with two columns, `sym` and the average expression; restrict the output to the first three genes.

Problem 472 Which gene has the largest average expression value in `normal_food` (`max`)? The smallest (`min`)? What is the average over the per gene average expression values (`avg`)?

Problem 473 Repeat these computations for `fatty_food`, that is, which genes have the largest and smallest average expression value? And what is the grand average expression value in `fatty_food`?

Problem 474 Next we compare `normal_food` and `fatty_food`. For this we need to `join` the two tables. Look at the first ten entries of the joined table:

```
select * from normal_food join fatty_food using (sym)
    limit 10;
```

Convert the join into a table with three columns: `sym`, the per gene average of `normal_food`, and the per gene average of `fatty_food`. Save the command in the file `join.sql`. Hint: In order to refer to attribute M1 from table `normal_food` as opposed to `fatty_food`, use the dot-notation as in

```
select normal_food.m1, fatty_food.m1
from ...
```

Problem 475 Our next aim is to compute the fold change between two average expression values. It is not straight forward to do this in SQL. So we leave sqlite (.quit) and enter on the command line

```
sqlite3 mouseExpress.db < join.sql |
tr '|' '\t'                        |
head
```

Problem 476 Instead of regenerating the joined table using join.sql and piping the result through tr, save this data in avg.txt for future reference.

7.2.3   Java

Problem 477 SQL is often embedded in a host language. Here is an example in Java for our database:

```
import java.sql.*;
public class MouseExpressDb{
    public static void main( String args[] ){
        String query = "select * from normal_food join
            fatty_food using(sym)";
        try{
            Class.forName("org.sqlite.JDBC");
            Connection c = DriverManager.getConnection
                ("jdbc:sqlite:mouseExpress.db");
            Statement stmt = c.createStatement();
            ResultSet rs = stmt.executeQuery(query);
            while(rs.next())
                    // Access column #1
                    System.out.println(rs.getString(1));
        }catch(Exception e){
            System.err.println(e.getClass().getName() +
                ": " + e.getMessage());
            System.exit(0);
        }
    }
}
```

Type this code into the file MouseExpressDb.java and compile it with

```
javac MouseExpressDb.java
```

and run the program

```
java -cp sqlite-jdbc-3.15.1.jar:. MouseExpressDb | head
```

where the `jar` file contains the JDBC-driver for `sqlite3`, which can be downloaded from the book web site. Next, copy `MouseExpressDb.java` to `MouseExpressDb2.java` and edit the code to search for the gene with the greatest difference between normal and fatty food.

7.2.4 ENSEMBL

Problem 478 To get a list of all databases hosted on the public ENSEMBL server, enter

```
mysql -h ensembldb.ensembl.org -u anonymous -e "show
   databases"
```

How many databases make up ENSEMBL?

Problem 479 We are interested in the databases for mouse (*Mus musculus*) and among those the `core` databases in particular. What is the version number of the latest `core` database for mouse?

Problem 480 The command

```
mysql -h ensembldb.ensembl.org -u anonymous -D
  someDatabase -e "show tables"
```

lists the tables of a particular database. How many tables make up the latest mouse `core` database?

Problem 481 To find out the attributes of a particular table, use

```
mysql ... -e "describe someTable"
```

Which attributes make up `seq_region`?

Problem 482 The following code returns the lengths of all mouse chromosomes

```
for a in $(seq 19) X Y
do
   mysql ... -e "select name,length from seq_region
      where coord_system_id = 3 and name = '$a'" |
   tail -n +2
done
```

What is the total length of the mouse genome?

Problem 483 What are the attributes of table `exon`?

Problem 484 Which fraction of the mouse genome is covered by exons?

Problem 485 Exons contained in all splice variants of a particular transcript are called "constitutive". Which fraction of the mouse genome is covered by constitutive exons?

Problem 486 Table `xref` contains the attribute `display_label`, which corresponds to the gene names we used in the expression analysis (`sym`). For each gene, `xref` also contains a `description` of its function. What is the function of the gene we found in Problem 475 with the greatest fold change in expressions?

Problem 487 Important entities in ENSEMBL such as genes or transcripts are labeled with a `stable_id`. A `stable_id` can, for example, be entered on the ENSEMBL web site (`www.ensembl.org`) to quickly and unambiguously look up information on a particular gene. We search for the `stable_id` that corresponds to *Hsd3b5* by joining `gene` and `xref` via `gene.display_xref_id` and `xref.xref_id`. What is the `gene.stable_id` of *Hsd3b5*?

Problem 488 Every gene has at least one, but possibly more, transcripts. They are listed in table `transcript`, which is connected to `gene` via the attribute `gene_id`. How many transcripts are known for *Hsd3b5*?

Chapter 8
Answers and Appendix: Unix Guide

8.1 Answers

Answer 1 You can either repeat `mkdir`

```
mkdir TestDir1
mkdir TestDir2
```

or write more succinctly

```
mkdir TestDir1 TestDir2
```

followed by

```
ls
```

Answer 2 The command

```
cd
```

changes to your "home directory". Try the command `pwd` to "print which directory" you are in.

Answer 3 The star, or "wildcard", matches any completion of `TestDir`. So

```
rmdir TestDir*
```

deletes `TestDir1` and `TestDir2`.

Answer 4 This causes the error message

```
rmdir: failed to remove TestDir1: Directory not empty
```

You can now either go into `TestDir1` and delete its contents or apply the "recursive" version of `rm`

```
rm -r TestDir1
```

Answer 5 Recall the wildcard character introduced in Problem 3:

```
rm *
```

The original version of the backmatter was revised: For detailed information please see Erratum. The erratum to this chapter is available at https://doi.org/10.1007/978-3-319-67395-0_9

© Springer International Publishing AG 2017
B. Haubold and A. Börsch-Haubold, *Bioinformatics for Evolutionary Biologists*, https://doi.org/10.1007/978-3-319-67395-0_8

Answer 6 Rename file a

```
mv a b
```

The command mv stands for "move". By moving a file from one name to another, it is renamed. To get more feedback on the action of mv, you can use -v to switch on its verbose mode.

Answer 7 The command history lists all previous commands, with history as the last one.

Answer 8 The shell prints

```
>
```

and waits for further input until the single quote in "can't" is closed. Alternatively, the command can be terminated using C-c twice. This is a very useful command for getting out of trouble.

Answer 9 The command goes into auto-repeat mode.

Answer 10 The command ls -t lists the files modified most recently first.

Answer 11 The help page contains a section on searching explaining that

```
/pattern
```

generates a search, and the next match can be found by pressing n.

Answer 12 Wc returns three numbers, the number of lines, words, and bytes in the input. In this case, the number of lines is identical to the number of words and both are equal to the number of files in the current directory.

Answer 13 Look up the bash man page

```
man bash
```

and search for pipe

```
/pipe
```

which gets you to the right place.

Answer 14 The command

```
x=1+1; echo $x
```

performs no addition, but simply returns the string 1+1.

Answer 15 The command

```
((x=1+1)); echo 'The result is $x'
```

prints the literal string rather than *interpolating* the variable $x as done with double quotes.

Answer 16 As with the double brackets, if let is dropped, x is assigned 1+1 as a string rather than as a calculation:

```
x=1+1; echo $x
1+1
```

Answer 17

```
let y=2**10; echo $y
1024
```

For mental arithmetic, it is useful to remember, $2^{10} \approx 10^3$.

Answer 18

```
let y=10-20; echo $y
-10
```

So, negative numbers work, too.

Answer 19

```
let y=10/3; echo $y
3
```

because the bash only computes with integers.

Answer 20

```
4^10
1048576
```

Alternatively, you could have estimated in your head

$$4^{10} = 2^{20} = \left(2^{10}\right)^2 \approx \left(10^3\right)^2 = 10^6.$$

Answer 21

```
echo 10/3 | bc
3
```

In other words, without -1, the default integer mode of bc is switched on.

Answer 22 List the contents of the root directory

```
ls /
```

Count the files and directories listed

```
ls / | wc -l
```

Answer 23 Change into BiProblems

```
cd BiProblems
```

List all files

```
ls -a
.  ..
```

This means the directory is empty. Nevertheless, there are two directories we can refer to. The first is denoted by one dot, which stands for the current directory, and the second by two dots, the parent directory in the tree.

Answer 24 Use

```
ls Data/ | wc -l
```

Answer 25 Count the number of FASTA files in `Data`:

```
ls Data/*.fasta | wc -l
```

Answer 26 Copy

```
cp ../Data/mgGenes.txt .
```

check

```
ls
```

and count the genes

```
wc -l mgGenes.txt
525 mgGenes.txt
```

Answer 27 The command `cat -n` adds numbers to the printed lines. For re-counting, we type

```
cat -n mgGenes.txt
```

and look at the last line of the output

```
525   MG_470   579224   580033    -
```

Answer 28 List the first ten lines:

```
head mgGenes.txt
MG_001   686     1828    +    dnaN
MG_002   1828    2760    +
MG_003   2845    4797    +    gyrB
MG_004   4812    7322    +    gyrA
MG_005   7294    8547    +    serS
MG_006   8551    9183    +    tmk
MG_007   9156    9920    +
MG_008   9923    11251   +
MG_009   11251   12039   +
MG_010   12068   12724   +
```

List the last ten lines:

```
tail mgGenes.txt
MG_462   566186   567640   -    gltX
MG_463   567627   568406   -
MG_464   568399   569556   -
MG_465   569528   569914   -    rnpA
MG_466   569883   570029   -    rpL34
MG_467   570055   570990   -
MG_468   570994   576345   -
MG_526   576351   577205   -
MG_469   577268   578581   -
MG_470   579224   580033   -
```

There are five columns: a key, start and end positions, the strand, and the gene symbol, if available.

Answer 29

```
grep + mgGenes.txt | wc -l
299
grep - mgGenes.txt | wc -l
227
```

The total number of genes (Problem 26) is 525, while $299 + 227 = 526$. One gene is counted twice, presumably because its name contains a "-".

Answer 30

```
cut -f 5 mgGenes.txt | grep -
rpmG-2
polC-2
```

Find the strand of *rpmG*

```
grep rpmG mgGenes.txt
MG_473   64367    64513     -     rpmG-2
MG_325   408793   408954    -     rpmG
```

and of *polC*

```
grep polC mgGenes.txt
MG_031   32359    36714     -     polC
MG_261   315701   318325    +     polC-2
```

So it was *polC-2* which resulted in the extra count for a gene on the minus strand.

Answer 31 Extract the genes

```
cut -f 1-4 mgGenes.txt | grep - > minus.txt
cut -f 1-4 mgGenes.txt | grep + > plus.txt
```

Check results

```
wc -l minus.txt
226 minus.txt

wc -l plus.txt
299 plus.txt
```

which add up to 525 genes. The simplest method to see that the genes on the plus strand do not form a homogeneous block is to enter

```
head -n 20 mgGenes.txt
```

which shows five genes on the minus strand. We can further quantify the mixing of genes between both strands: If the genes on the plus strand did form a homogeneous block, the first 299 genes in mgGenes.txt would all be on the plus strand. So we look at the first 299 genes in mgGenes.txt and count how many of them are on the plus strand:

```
head -n 299 mgGenes.txt | grep + | wc -l
246
```

Hence, there are only $299 - 246 = 53$ genes on the minus strand among the first 246 genes on the plus strand.

Answer 32 Leftward redirection (<) writes the contents of a file to the standard input stream, stdin, from where it can be read by any tool, for example, `grep`:

```
grep + < mgGenes.txt
```

Answer 33

```
man sudo
```

explains that `sudo` switches into sys-admin mode.

Answer 34 Without the ampersand, the command line freezes until the child window running `emacs` is closed again.

Answer 35 Saving to `mgGenes2.txt` can be done in `emacs` via the menu. Searching for `polC-2` can also be done using the menu. Now check the difference between `mgGenes.txt` and `mgGenes2.txt`:

```
diff mgGenes.txt mgGenes2.txt
56c56
< MG_473          64367    64513    -       rpmG-2
---
> MG_473          64367    64513    -       rpmG_2
288c288
< MG_261          315701   318325   +       polC-2
---
> MG_261          315701   318325   +       polC_2
```

This means line 56 of `mgGenes.txt` was changed (c) into line 56 of `mgGenes2.txt` and similarly for line 288.

Answer 36 Extract, for example, line 56 of `mgGenes.txt`:

```
head -n 56 mgGenes.txt | tail -n 1
```

Answer 37 The one exception is C-w, which deletes the word to the left of the cursor in `bash`, but deletes a selected region in `emacs`.

Answer 38 Exit emacs with

```
C-x C-c
```

Notice the file `mgGenes2.txt`, which is a backup file created by `emacs` when you worked on `mgGenes2.txt`. You can ignore it for now, but if in some later `emacs` session something disastrous happens to the original file, you can always go to the backup.

Answer 39 The command is not found:

```
drawGenes
drawGenes: command not found
```

Answer 40 On Ubuntu, we get

```
which ls
/bin/ls
```

Answer 41

```
find ~/ -name ".bashrc"
```

Answer 42 Draw the figure with

```
drawGenes exampleGenes.txt |
gnuplot -p -e 'unset ytics; plot[][-10:10] "< cat" title ""
    with lines'
```

If `gnuplot` is not installed on your system, you need to install it at this point. Notice also that we have distributed the commands of the pipeline over two lines to make them easier to read. You can enter them either all on the same line or on separate lines as shown.

Answer 43 The shortest version we found was

```
drawGenes exampleGenes.txt |
gnuplot -p -e 'uns yti; p[][-10:10] "< cat" t "" w l'
```

Answer 44 The command `set xlabel` can be abbreviated to `se xl`:

```
drawGenes exampleGenes.txt |
gnuplot -p -e 'se xl "Position (bp)"; uns yti; p[][-10:10]
    "< cat" t "" w l'
```

Answer 45 Draw the genes:

```
cut -f 2-4 mgGenes.txt |
drawGenes              |
gnuplot -p -e 'se xl "Position (bp)"; unset ytics; plot
    [][-10:10] "<cat " t "" w l'
```

to get

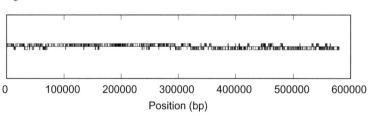

On the 5'-half of the genome, the genes are predominantly on the forward strand, on the 3'-half predominantly on the reverse strand.

Answer 46

```
./drawGenes.sh
bash: ./drawGenes.sh: Permission denied
```

Answer 47

```
drawGenes.sh
drawGenes.sh: No such file or directory
```

Answer 48 As before, we reset the environment

```
source ~/.bashrc
```

and should then be able to execute drawGenes.sh.

Answer 49 The preliminaries were

```
cd                 # change into the home directory
cd BiProblems
mkdir UnixScripts
cd UnixScripts
```

As to "Hello World!",

```
echo 'Hello   World!'
```

or

```
echo Hello'   'World!
```

Answer 50 When executing

```
for((i=1; i<=10; i++)); do echo -n 'Hello World!'; done
```

the newline usually added by echo is omitted. This is useful for printing more than one item on the same line.

Answer 51

```
for i in $(seq 10)
do
    echo $i
done
```

Answer 52

```
for i in $(seq 10)
do
    echo -n $i ' '
done
```

Note the blank in single quotes at the end of the echo command to separate the numbers. This might also have been surrounded by double quotes. However, recall from Problem 15 that interpolation of variable values only works inside double quotes:

```
for i in $(seq 10)
do
    echo -n "$i "
done
```

Answer 53 The command

```
man seq
```

tells us how to count in steps of two

```
seq 0 2 10
```

or backward

```
seq 10 -2 0
```

Answer 54

```
for i in $(seq 525)
do
    cut -f 4 mgGenes.txt |
        head -n $i       |
        grep +           |
        wc -l
done
```

Answer 55 All that is missing is a command for printing the number of genes analyzed:

```
for i in $(seq 525)
do
    echo -n $i ' '           # print number of genes analyzed
    cut -f 4 mgGenes.txt |
head -n $i               |
grep +                   |
wc -l
done
```

execute this

```
bash countGenes.sh |
gnuplot -p -e 'se xl "NumGenes"; se yl "NumPlus"; p "< cat "
    t "" w l'
```

to get the number of *M. genitalium* genes on the plus strand as a function of the number of genes.

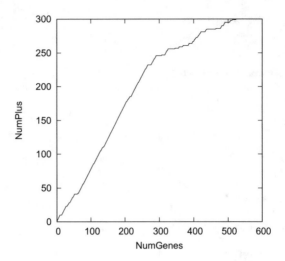

Answer 56 The script is

```
for strand in + -
do
      for i in $(seq 525)
      do
            echo -n $i ' '
            cut -f 4 mgGenes.txt|
            head -n $i            |
            grep $strand          |
            wc -l
      done
      echo ' '
done
```

Notice the command

```
echo ' '
```

in the penultimate line, which separates the two data sets by a blank line. Run this

```
bash countGenes.sh |
gnuplot -p -e 'se xl "NumGenes"; se yl "NumPlus"; p "< cat "
      t "" w l'
```

to get

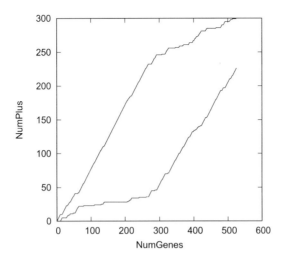

The blank line in the data set causes gnuplot to draw a second graph in the same plot window, which is useful for getting a first impression of the data. However, it does not use two plot symbols for distinguishing between the + and the – data. We learn how to do this later when we have gained more experience with gnuplot.

Answer 57 Carry out the substitution:

```
sed 's/-2/_2/' mgGenes.txt > mgGenes3.txt
```

and check its result

```
diff mgGenes2.txt mgGenes3.txt
```

No differences should be found.

Answer 58

```
cut -f 5 mgGenes.txt | sed '/^$/d' | wc -l
219
```

Answer 59 Save the edited file in a new file name

```
sed '56p' mgGenes3.txt > mgGenes4.txt
```

and compare the two files

```
diff mgGenes3.txt mgGenes4.txt
56a57
> MG_473          64367    64513      -      rpmG_2
```

To find out what this means, look at the two lines mentioned, 56 and 57, in both files:

```
sed -n '56p' mgGenes3.txt
MG_473   64367   64513       -      rpmG_2
sed -n '57p' mgGenes3.txt
MG_474   64527   64910       -
```

and

```
sed -n '56p' mgGenes4.txt
MG_473   64367   64513      -      rpmG_2
sed -n '57p' mgGenes4.txt
MG_473   64367   64513      -      rpmG_2
```

In other words, line 56 in mgGenes4.txt was appended to form line 57, which was summarized by diff as 56a57.

Answer 60

```
sed -n '1,10p' mgGenes3.txt > t1.txt
head mgGenes3.txt > t2.txt
diff t1.txt t2.txt
```

Answer 61 The original command was

```
cut -f 5 mgGenes3.txt | grep -
```

Its sed version is

```
cut -f 5 mgGenes3.txt | sed -n '/-/p'
```

Answer 62 Smallest position:

```
cut -f 2,3 mgGenes3.txt | sed 's/\t/\n/' | sort -n | head -n
     1
```

Largest position:

```
cut -f 2,3 mgGenes3.txt | sed 's/\t/\n/' | sort -n | tail -n
     1
```

Answer 63 By accidentally omitting -n from sort, the smallest gene position would seem to be

```
cut -f 2,3 mgGenes3.txt | sed 's/\t/\n/' | sort | head -n 1
102454
```

Answer 64 Cut out the gene positions as given:

```
cut -f 2,3 mgGenes3.txt > t1.txt
```

Cut out the gene positions and sort them:

```
cut -f 2,3 mgGenes3.txt | sort -n > t2.txt
```

Check that the two files are identical

```
diff t1.txt t2.txt
```

Answer 65 Overlapping genes induce a list of positions that is not strictly ascending; in our example, this is

```
1000
2000
1990
3000
```

To find out whether any genes overlap in *M. genitalium*, we need to compare the sorted and the unsorted list of positions:

```
cut -f 2,3 mgGenes3.txt | sed 's/\t/\n/' | sort -n > t1.txt
cut -f 2,3 mgGenes3.txt | sed 's/\t/\n/' > t2.txt
diff t1.txt t2.txt
```

This returns a long list of differences. Overlapping genes appear to be quite common in *M. genitalium*.

Answer 66 There are many ways to count the gyrases; one example is to remove all other genes:

```
cut -f 5 mgGenes3.txt | sed -n -f filter.sed
gyrB
gyrA
```

where `filter.sed` contains a single line:

```
/gyr/p    # print gyrases
```

Answer 67

```
awk '{print $5}' mgGenes3.txt
```

Notice that the code inside the curly brackets is executed for every input line.

Answer 68 Enter

```
awk '{print "Sum: " $1 + $2}'
```

Then enter two numbers separated by a blank to get their sum. Press C-c twice to get out of this infinite loop.

Answer 69 Print genes on the plus strand

```
awk '$4~/[+]/{print}' mgGenes3.txt
```

and on the minus strand

```
awk '$4~/[-]/{print}' mgGenes3.txt
```

Notice that without argument, `print` prints the entire input line, which is equivalent to

```
print $0
```

Answer 70 Without the action block, each matching line is printed, for example

```
awk '$4~/[-]/' mgGenes3.txt
```

Answer 71 Use

```
awk -f drawGenes.awk mgGenes3.txt |
gnuplot -p pipe.gp
```

where `drawGenes.awk` is

```
$4~/[+]/{
    x = 1
}
$4~/[-]/{
    x = -1
}
{
    print $2 "\t" 0
    print $2 "\t" x
    print $3 "\t" x
    print $3 "\t" 0
}
```

to get the familiar plot of *M. genitalium* genes:

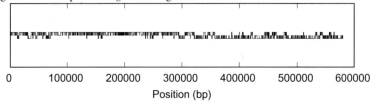

Answer 72 Find the shortest gene:

```
awk '{print $1 "\t" $3-$2+1}' mgGenes3.txt | sort -k 2 -n |
    head
MG_479    74
MG_489    74
MG_493    74
MG_496    74
MG_499    74
MG_495    75
MG_501    75
MG_502    75
MG_504    75
MG_512    75
```

So there are five genes of length 74. Find the longest gene:

```
awk '{print $1 "\t" $3-$2+1}' mgGenes3.txt | sort -k 2 -n |
    tail
MG_309    3678
MG_338    3813
MG_340    3879
MG_064    3996
MG_341    4173
MG_191    4335
```

```
MG_031    4356
MG_386    4851
MG_468    5352
MG_218    5418
```

There is a single gene of length 5418. Notice the option "-k 2" directing sort to use entries in the second column as sort keys.

Answer 73

```
awk '{s += $3-$2+1; c++}END{print "Avg: " s/c}' mgGenes3.txt
Avg: 1029.42
```

If, by any chance, you wrote = instead of +=, the program prints the length of the last gene divided by 525, approximately 1.54.

Answer 74 The program to compute the variance:

```
{
    l = $3 - $2 + 1        # length
    len[n++] = l           # store
}END{
    for(i=0; i<n; i++)     # mean
        m += len[i]
    m /= n
    for(i=0; i<n; i++) {   # variance
        x = m - len[i]
        s += x * x
    }
    printf "Var:  %e\n", s / (n - 1)
}
```

Notice that the mean for loop has no curly brackets, while the variance for loop does. A for loop with a single line as action block needs no curly brackets, though they would not cause an error either. An action block consisting of more than one line needs to be delineated by curly brackets. This rule equally applies to if. When run on the gene lengths of *M. genitalium*, var.awk produces

```
awk -f var.awk mgGenes3.txt
Var:  6.623986e+05
```

The program var calculates the same, very large, variance:

```
awk '{print $3-$2+1}' mgGenes3.txt | var
#mean    var
1.029423e+03   6.623986e+05
```

Answer 75 Here is the program.

```
{
    l = $3-$2+1
    len[n++] = l
    sum += l
}END{
```

```
for(i=0; i<n; i++){      # observed
    s += len[i] / sum
    print i "\t" s
}
print ""
for(i=0; i<n; i++)       # expected
    print i "\t" i/n
}
```

Piping its result through

```
gnuplot -p -e 'set xlabel "Rank"; set ylabel "Cumulative
    Length"; plot
"< cat " title "" wi li'
```

gives

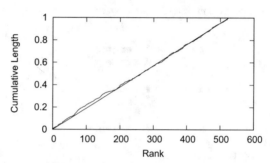

So there is no systematic bias with respect to gene length along the genome.

Answer 76 The number of names used

```
cut -f 5 mgGenes3.txt | sed '/^$/d' | wc -1
219
```

is equal to the number of unique names:

```
cut -f 5 mgGenes3.txt | sed '/^$/d' | sort | uniq | wc -1
219
```

So all names are unique.

Answer 77

```
awk '{print $3-$2+1}' mgGenes3.txt | sort | uniq | wc -1
365
```

Answer 78

```
awk '{print $3-$2+1}' mgGenes3.txt | sort | uniq -c | sort -
    n -r | head -n 5
    9 77
    7 76
    6 75
    5 74
    4 936
```

Answer 79 Generate the output with `uniq -c`

```
awk '{print $3-$2+1}' mgGenes3.txt  |
sort                                |
uniq -c                             |
sed 's/^ *//'                       |
sort > t1.txt
```

The second `sort` is necessary to achieve sorting according to the count printed by `uniq`. Now reproduce this result with `uniqC.awk`

```
awk '{print $3-$2+1} ' mgGenes3.txt |
awk -f uniqC.awk                    |
sort > t2.txt
```

Check the two files are identical

```
diff t1.txt t2.txt
```

Answer 80

```
awk '{print $3-$2+1} ' mgGenes3.txt        |
awk -f uniqC.awk                           |
awk '{print $2 "\t" $1}'                   |
sort -n                                    |
gnuplot -p -e 'set xlabel "Length"; set ylabel "Count"; plot
     "< cat" title "" wi li'
```

gives

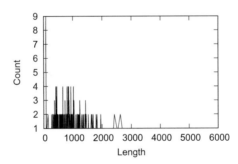

Answer 81 Replace the 10 in the `for` loop by `seqLen`

```
BEGIN{
    print ">Random_Sequence"
    srand(seed)
    s[0] = "A"; s[1] = "T"; s[2] = "C"; s[3] = "G"
    for(i=0; i<seqLen; i++){
        j = int(rand() * 4)
        printf("%s", s[j])
    }
    printf("\n")
}
```

and run

```
awk -v seed=$RANDOM -v seqLen=30 -f ranSeq.awk
```

Answer 82 Here is one possible answer:

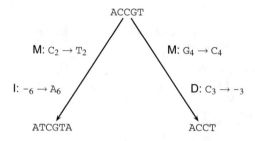

where M is the mutation, I is the insertion, and D is the deletion. You might have been tempted to mark the deletion event as a gap in the final sequence, but remember that sequences exist as gapless molecules. The three types of evolutionary events marked in the above graph, mutation, inversion, and deletion, only become visible by sequence comparison, that is, in an alignment:

$$
\begin{array}{l}
\text{ATCGTA} \\
\text{AC-CT-}
\end{array}
$$

Answer 83 Here is one solution out of many:

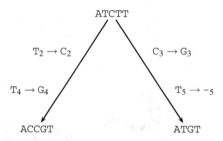

Answer 84 Here are five possible answers, you might well have found others:

ACCGT	ACCGT-	ACCGT
-ATGT	-A-TGT	A-TGT
Events = 3	Events = 6	Events = 2
ACCGT	ACCGT-	
AT-GT	AT-GT	
Events = 2	Events = 5	

Answer 85 The re-scored results look like this:

```
ACCGT              ACCGT-             ACCGT
-ATGT              -A-TGT             A-TGT
Events = 3         Events = 6         Events = 2
Score = -11        Score = -30        Score = -7

ACCGT              ACCGT-
AT-GT              AT-GT
Events = 2         Events = 5
Score = -7         Score = -21
```

Answer 86 This is expressed by

$$g = g_o + (l - 1) \times g_e.$$

Answer 87 Read man echo to learn that -e enables the interpretation of "unprintable" characters such as a newline:

```
echo -e '>Seq1\nACCGT'
>Seq1
ACCGT
```

And saved:

```
echo -e '>Seq1\nACCGT' > seq1.fasta
```

Without -e:

```
echo '>Seq1\nACCGT'
>Seq1\nACCGT
```

Answer 88 The command

```
gal -i seq1.fasta -j seq2.fasta
```

returns

```
Query: 1 ACCGT 5
         |  ||
Sbjct: 1 A-TGT 4
```

with a score of -7. Since by default gap opening (g_o) is -5 and gap extension (g_e) is -2, the gap scoring scheme is

$$g = g_o + l \times g_e.$$

Another alignment with the same score is

```
Query: 1 ACCGT 5
         |  ||
Sbjct: 1 A-TGT 4
```

Answer 89 Copy hbb1.fasta and hbb2.fasta

```
cp ../Data/hbb1.fasta .
cp ../Data/hbb2.fasta .
```

Note the dot at the end of the command, which means copy into the current directory. Look up the function of hbb1.fasta

```
head -n 1 hbb1.fasta
>gi|28302128|ref|NM_000518.4| Homo sapiens hemoglobin, beta
    (HBB), mRNA
```

and of hbb2.fasta

```
head -n 1 hbb2.fasta
>gi|6003531|gb|AF181832.1|AF181832 Homo sapiens hemoglobin
    beta subunit variant (HBB) mRNA, partial cds
```

So the first sequence is the mRNA of human beta hemoglobin, and the second a partial coding sequence (CDS) of the same protein. Run gal

```
gal -i hbb1.fasta -j hbb2.fasta
```

to find these two DNA sequences differ by large gaps at the 5' and 3' ends, and a single mismatch in between.

Answer 90 Running gal with default options shows that the mismatch is close to position 400. So rerunning gal with -l 100 lets you conveniently find that the mutation is in position 397 in hbb1.fasta and 290 in hbb2.fasta.

Answer 91 The mutation is in position 290 of hbb2.fasta, which is translated in frame. The position of the mutation within the codon can thus be found as the remainder when dividing the position by 3, the modulo operation, 290 mod 3 = 2. Hence, the mutation affects the two codons

```
GCC
GAC
```

which encode alanine and aspartate. The mutation is non-synonymous.

Answer 92 The probability of finding a stop codon in a random sequence is 3/64, so given a start codon, the expected distance to the next stop codons is $64/3 \approx 21$ amino acids. This is the expected length of open reading frames in random DNA sequences.

Answer 93 Compute the average ORF length:

```
awk -v seed=$RANDOM -v n=10000 -f simOrf.awk |
awk '{s+=$1;c++}END{print s/c}'
21.0919
```

This is close to the expectation of 21.

Answer 94

```
awk -v seed=$RANDOM -v n=1000 -f simOrf.awk |
histogram                                    |
gnuplot -p orf.gp
```

where orf.gp is

```
set xlabel "Length"
set ylabel "Frequency"
plot "< cat" title "" with lines
```

to get

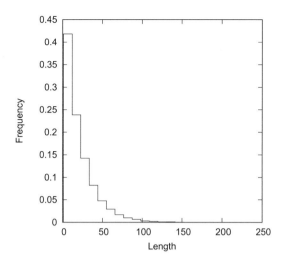

Answer 95 The two most polar amino acids are aspartate and glutamate.

Answer 96 Phe→Leu: 1; Phe→Trp: 2; Phe→Glu: 3.

Answer 97 Set up the session:

```
mkdir AminoAcidMat
cd AminoAcidMat
cp ../Data/polarity.dat .
```

Find the least polar amino acid:

```
sed '/#/d' polarity.dat | sort -k 2 -n | head -n 1
Cys      4.8
```

The most polar:

```
sed '/#/d' polarity.dat | sort -k 2 -n | tail -n 1
Asp        13.0
```

Answer 98 Smallest: 1, M and W; largest: 6, L, S, and R.

Answer 99 There are two examples at the first codon position: TTA and TTG encode L, and so do CTA/CTG. There are no synonymous mutations at the second codon position.

Answer 100 The number of possible arrangements for n books on a shelf is n factorial

$$n(n - 1)(n - 2)...2 = n!$$

Compute

```
awk 'BEGIN{p = 1; for(i=2;i<=20;i++)p *= i; printf("%e\n", p
    )}'
2.432902e+18
```

to find $n! \approx 2.4 \times 10^{18}$.

Answer 101 The three possible changes at the second position yield serine (S), tyrosine (Y), and cysteine (C). Changes at the third position again yield leucine (L). With the exception of serine, these mutant amino acids also have a similar polarity as phenylalanine.

Answer 102 Execute

```
genCode -p polarity.dat   |
cut -f 2                   |
histogram                  |
gnuplot -p ms0.gp
```

where ms0.gp is

```
set ylabel "Frequency"
set xlabel "MS_0"
set arrow from 5.19,0.02 to 5.19,0
plot "< cat" title "" with lines
```

This gives

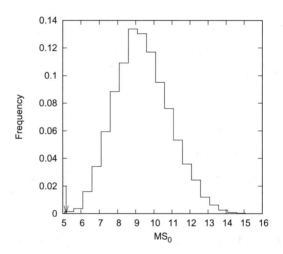

The arrow indicates the MS_0 value of the natural code, 5.19, which is quite untypical when compared to the MS_0 values from random codes. Notice that this is the result of a simulation—yours is bound to differ slightly.

Answer 103 We got between 0 and 2 better codes. Since genCode goes through 10^4 random codes, the proportion of better codes is about 10^{-4}.

Answer 104 Run genCode with, say, 10^6 iterations

```
genCode -n 1000000 polarity.dat | grep ms0: | wc -l
117
```

When rejecting the null hypothesis that the natural code is not mutation-optimized, the error probability is $P = 117 \times 10^{-6} \approx 1.2 \times 10^{-4}$, so we reject and conclude that the code is indeed significantly optimized with respect to polarity.

Answer 105 Run commands like

```
genCode -n 1000000 hydropathy.dat   | grep ms0:  | wc -l
```

to find

Attribute	P
Polarity	1.2×10^{-4}
Hydropathy	7.8×10^{-3}
Volume	0.11
Charge	0.56

Answer 106 A→S: 0.0028; S→A: 0.0035.

Answer 107

```
tail -n 20 pam1.txt  | awk '{c++; s+=$(c+1)*0.05}END{print
   (1-s)*100}'
0.977
```

In other words, there is approximately a one percent probability of finding a mismatched amino acid, as would be expected for 1 PAM.

Answer 108 Compute M^{100}

```
pamPower -n 100 pam1.txt
```

to get (edited for legibility)

	A	R	N	D	C	Q	E	G	H	I	L	K	M	F	P	S	T	W	Y	V
A	**.32**	.03	.07	.07	.03	.05	.08	.10	.03	.05	.03	.03	.04	.02	.11	.13	.13	.01	.02	.08
R	.01	**.44**	.02	.01	.01	.05	.01	.01	.05	.02	.01	.09	.03	.01	.03	.03	.02	.05	.00	.01
N	.03	.02	**.21**	.10	.01	.03	.05	.03	.07	.02	.01	.05	.01	.01	.02	.06	.04	.01	.02	.01
D	.04	.01	.11	**.31**	.00	.05	.16	.04	.04	.01	.01	.03	.01	.00	.02	.04	.03	.00	.01	.01
C	.01	.01	.01	.00	**.77**	.00	.00	.00	.01	.01	.00	.00	.01	.00	.00	.02	.01	.00	.02	.01
Q	.02	.04	.03	.04	.00	**.32**	.09	.01	.09	.01	.02	.03	.02	.00	.03	.02	.02	.00	.00	.01
E	.05	.01	.06	.17	.00	.12	**.32**	.03	.03	.02	.01	.03	.01	.00	.03	.03	.03	.00	.01	.02
G	.11	.02	.07	.07	.02	.03	.06	**.55**	.02	.02	.01	.03	.02	.01	.04	.10	.05	.00	.01	.04
H	.01	.04	.06	.03	.01	.08	.02	.01	**.43**	.01	.01	.02	.01	.01	.02	.01	.01	.01	.02	.01
I	.02	.01	.02	.01	.01	.01	.01	.01	.01	**.32**	.05	.01	.06	.04	.01	.01	.03	.00	.01	.12
L	.03	.02	.02	.01	.00	.04	.01	.01	.03	.12	**.61**	.02	.21	.09	.02	.02	.03	.03	.03	.10
K	.03	.18	.10	.05	.01	.07	.05	.02	.04	.03	.02	**.51**	.09	.01	.03	.05	.06	.01	.01	.02
M	.01	.01	.00	.00	.00	.01	.00	.00	.00	.02	.04	.02	**.29**	.01	.00	.01	.01	.00	.00	.02
F	.01	.01	.01	.00	.00	.00	.00	.01	.02	.04	.04	.00	.03	**.60**	.00	.01	.01	.02	.17	.01
P	.06	.03	.02	.02	.01	.04	.03	.02	.03	.01	.02	.02	.01	.01	**.49**	.06	.04	.00	.00	.02
S	.10	.05	.10	.05	.05	.04	.05	.08	.03	.03	.01	.05	.03	.02	.08	**.25**	.12	.03	.02	.03
T	.09	.03	.06	.04	.02	.03	.03	.03	.02	.05	.02	.05	.04	.01	.04	.10	**.31**	.01	.02	.05
W	.00	.01	.00	.00	.00	.00	.00	.00	.00	.00	.00	.00	.00	.01	.00	.00	.00	**.79**	.01	.00
Y	.01	.00	.01	.00	.02	.00	.01	.00	.02	.01	.01	.00	.01	.13	.00	.01	.01	.02	**.59**	.01
V	.06	.02	.02	.02	.02	.02	.02	.02	.02	.21	.07	.02	.09	.02	.03	.03	.06	.00	.02	**.42**

Compute M^{1000}

```
pamPower -n 1000 pam1.txt
```

to get

	A	R	N	D	C	Q	E	G	H	I	L	K	M	F	P	S	T	W	Y	V
A	**.09**	.09	.09	.09	.09	.09	.09	.09	.09	.09	.09	.09	.09	.09	.09	.09	.09	.07	.08	.09
R	.04	**.04**	.04	.04	.04	.04	.04	.04	.04	.04	.04	.04	.04	.04	.04	.04	.04	.05	.04	.04
N	.04	.04	**.04**	.04	.04	.04	.04	.04	.04	.04	.04	.04	.04	.04	.04	.04	.04	.04	.04	.04
D	.05	.05	.05	**.05**	.04	.05	.05	.05	.05	.05	.05	.05	.05	.04	.05	.05	.05	.04	.04	.05
C	.03	.03	.03	.03	**.09**	.03	.03	.03	.03	.03	.03	.03	.03	.03	.03	.03	.03	.02	.03	.03
Q	.04	.04	.04	.04	.03	**.04**	.04	.04	.04	.04	.04	.04	.04	.04	.04	.04	.04	.03	.03	.04
E	.05	.05	.05	.05	.05	.05	**.05**	.05	.05	.05	.05	.05	.05	.05	.05	.05	.05	.04	.04	.05
G	.09	.09	.09	.10	.09	.09	.10	**.10**	.09	.09	.09	.10	.09	.08	.10	.09	.10	.07	.08	.09
H	.03	.03	.03	.03	.03	.03	.03	.03	**.03**	.03	.03	.03	.03	.03	.03	.03	.03	.03	.03	.03
I	.04	.04	.04	.04	.04	.04	.04	.04	.04	**.04**	.04	.04	.04	.04	.04	.04	.04	.03	.04	.04
L	.09	.08	.08	.08	.08	.08	.08	.08	.08	.09	**.10**	.09	.09	.10	.09	.08	.09	.09	.09	.09
K	.08	.09	.09	.09	.07	.09	.09	.08	.08	.08	.08	**.09**	.08	.08	.09	.08	.09	.08	.07	.08
M	.02	.01	.02	.01	.01	.02	.01	.01	.01	.02	.02	.02	**.02**	.02	.02	.02	.02	.01	.02	.02
F	.04	.04	.04	.04	.04	.04	.04	.04	.04	.05	.05	.04	.05	**.07**	.04	.04	.04	.05	.07	.04
P	.06	.05	.06	.06	.05	.06	.06	.06	.05	.05	.05	.06	.05	.05	**.06**	.06	.06	.05	.05	.05
S	.07	.07	.07	.07	.07	.07	.07	.07	.07	.07	.07	.07	.07	.07	.07	**.07**	.07	.06	.06	.07
T	.06	.06	.06	.06	.06	.06	.06	.06	.06	.06	.06	.06	.06	.06	.06	.06	**.06**	.05	.05	.06
W	.01	.01	.01	.01	.01	.01	.01	.01	.01	.01	.01	.01	.01	.01	.01	.01	.01	**.10**	.01	.01
Y	.03	.03	.03	.03	.03	.03	.03	.03	.03	.03	.03	.03	.03	.05	.03	.03	.03	.04	**.05**	.03
V	.06	.06	.06	.06	.06	.06	.06	.06	.06	.07	.07	.07	.07	.07	.07	.06	.07	.06	.06	**.07**

which has much more uniform values within a row than M^{100}.

Answer 109 Here is the computation:

```
for a in 1 2 5 10 20 50 100 200 500 1000
do
    echo -n $a ' '
    pamPower -n $a pam1.txt        |
      tail -n +2                   |
      awk '{s+=$(NR+1)}END{print (1-s/20)*100}'
done
```

Visualize the result

```
bash pamPower.sh   |
gnuplot -p pamPower.gp
```

where pamPower.gp is

```
set xlabel "PAM"
set ylabel "%-Difference"
plot "< cat" title "" with lines
```

to get

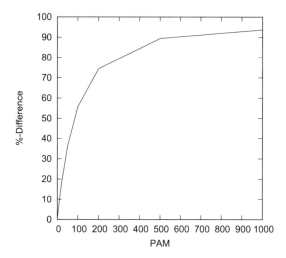

Answer 110 To make it easier to connect the sorted frequencies with amino acids, we printed the amino acids in front of the frequencies:

```
awk -f labelFreq.awk aa.txt | sort -k 2 -n
```

where `labelFreq.awk` is

```
/^#/{
    # Store amino acid designations
    for(i=2; i<=21; i++)
        aa[i-1] = $i
}
!/^#/{
    c++   # Count lines
    print aa[c] "\t" $1
}
```

This tells us that the least frequent amino acid is tryptophane (W), the most frequent glycine (G).

Answer 111 The amino acid frequencies

A	R	N	D	C	Q	E	G	H	I	L	K	M	F	P	S	T	W	Y	V
.09	.04	.04	.05	.03	.04	.05	.09	.03	.04	.09	.08	.01	.04	.05	.07	.06	.01	.03	.07

are similar to the columns in M^{1000}. In other words, after 1000 PAM of evolutionary time, homologous proteins contain pairs of amino acids at frequencies similar to the frequencies in random sequence pairs.

Answer 112 Here is the pipeline:

```
pamPower -n 70 pam1.txt |
pamNormalize -a aa.txt
```

Observe that normalization makes the matrix symmetrical, which is what we need when scoring pairwise alignments, where we cannot tell the direction of change.

Answer 113 Here is the final pipeline

```
pamPower -n 70 pam1.txt   |
pamNormalize -a aa.txt    |
pamLog > pam70sm.txt
```

This pipeline generates the following matrix (typeset to improve readability):

	A	R	N	D	C	Q	E	G	H	I	L	K	M	F	P	S	T	W	Y	V
A	5	-4	-1	-1	-4	-2	-1	0	-4	-2	-4	-4	-3	-5	0	1	1	-9	-5	-1
R	-4	8	-3	-6	-5	0	-5	-6	0	-3	-6	2	-2	-6	-2	-1	-4	0	-8	-5
N	-1	-3	6	3	-7	-1	0	-1	1	-3	-5	0	-5	-5	-3	1	0	-7	-3	-4
D	-1	-6	3	6	-9	0	3	-1	-2	-5	-8	-2	-7	-9	-4	-1	-2	-11	-7	-5
C	-4	-5	-7	-9	9	-9	-9	-7	-5	-4	-10	-9	-6	-8	-6	-1	-4	-11	-2	-4
Q	-2	0	-1	0	-9	7	2	-4	2	-5	-3	-1	-2	-8	-1	-3	-3	-9	-8	-4
E	-1	-5	0	3	-9	2	6	-2	-2	-4	-6	-2	-5	-9	-3	-2	-3	-11	-6	-4
G	0	-6	-1	-1	-7	-4	-2	6	-6	-6	-7	-5	-7	-6	-3	0	-3	-10	-9	-3
H	-4	0	1	-2	-5	2	-2	-6	8	-6	-4	-3	-6	-3	-2	-3	-4	-6	-1	-5
I	-2	-3	-3	-5	-4	-5	-4	-6	-6	7	0	-4	1	-1	-5	-4	-1	-10	-4	3
L	-4	-6	-5	-8	-10	-3	-6	-7	-4	0	6	-5	2	-1	-4	-6	-4	-6	-4	0
K	-4	2	0	-2	-9	-1	-2	-5	-3	-4	-5	6	0	-9	-4	-2	-1	-7	-8	-5
M	-3	-2	-5	-7	-6	-2	-5	-7	-6	1	2	0	10	-2	-5	-3	-2	-9	-7	0
F	-5	-6	-5	-9	-8	-8	-9	-6	-3	-1	-1	-9	-2	8	-7	-4	-5	-2	4	-5
P	0	-2	-3	-4	-6	-1	-3	-3	-2	-5	-4	-4	-5	-7	7	0	-2	-9	-9	-4
S	1	-1	1	-1	-1	-3	-2	0	-3	-4	-6	-2	-3	-4	0	5	2	-3	-5	-3
T	1	-4	0	-2	-4	-3	-3	-3	-4	-1	-4	-1	-2	-5	-2	2	6	-8	-5	-1
W	-9	0	-7	-11	-11	-9	-11	-10	-6	-10	-6	-7	-9	-2	-9	-3	-8	13	-2	-11
Y	-5	-8	-3	-7	-2	-8	-6	-9	-1	-4	-4	-8	-7	4	-9	-5	-5	-2	9	-5
V	-1	-5	-4	-5	-4	-4	-4	-3	-5	3	0	-5	0	-5	-4	-3	-1	-11	-5	6

The score of the alignment is accordingly $1 + 0 + 6 + 5 + 3 = 15$.

Answer 114 There are 14 pairs of mismatched amino acids with a positive score:

#	Amino acid pair	Score
1	AS	1
2	AT	1
3	RK	2
4	ND	3
5	NH	1
6	NS	1
7	DE	3
8	QE	2
9	QH	2
10	IM	1
11	IV	3
12	LM	2
13	FY	4
14	ST	2

Such residue pairs tend to have similar structures and polarities. Phenylalanine (F) and tyrosine (Y) have the largest positive score. Both amino acids have similar shapes (Fig. 2.3) and polarities (Fig. 2.4).

Answer 115 The correct reading frame contains a long stretch of amino acids without stop codons (*):

```
transeq -frame 3 -filter < hbb1.fasta > hbb1prot.fasta
transeq -frame 1 -filter < hbb2.fasta > hbb2prot.fasta
```

Align the sequences

```
gal -i hbb1prot.fasta -j hbb2prot.fasta -p -m pam70sm.txt
```

to get

```
Query:              >NM_000518.4_3 Homo sapiens hemoglobin, beta (HBB),
   mRNA
  Length:           208
Subject:            >AF181832_1 Homo sapiens hemoglobin beta subunit
                    variant (HBB) mRNA, partial cds
  Length:           163
Score:              964.0


Q: 1   ICF*HNCVH*QPQTDTMVHLTPEEKSAVTALWGKVNVDEVGGEALGRLLVVYPWTQRFFE 58
                                      NVDEVGGEALGRLLVVYPWTQRFFE
S: 1   ------------------------------NVDEVGGEALGRLLVVYPWTQRFFE 25


Q: 59  SFGDLSTPDAVMGNPKVKAHGKKVLGAFSDGLAHLDNLKGTFATLSELHCDKLHVDPENF 118
       SFGDLSTPDAVMGNPKVKAHGKKVLGAFSDGLAHLDNLKGTFATLSELHCDKLHVDPENF
S: 26  SFGDLSTPDAVMGNPKVKAHGKKVLGAFSDGLAHLDNLKGTFATLSELHCDKLHVDPENF 85


Q: 119 RLLGNVLVCVLAHHFGKEFTPPVQAAYQKVVAGVANALAHKYH*ARFLAVQFLLKVPLFP 177
       RLLGNVLVCVL HHFGKEFTPPVQAAYQKVVAGVANALAHKYH*ARFLAVQFLLKVPLFP
S: 86  RLLGNVLVCVLDHHFGKEFTPPVQAAYQKVVAGVANALAHKYH*ARFLAVQFLLKVPLFP 144


Q: 178 KSNY*TGGYYEGP*ASGFCLIKNIYFHC 203
       KSNY*TGGYYEGP*ASG
S: 145 KSNY*TGGYYEGP*ASG----------X 160
//
```

The row between query and subject indicates the match state: the residue for an exact match and a blank for a mismatch with a score of zero or less. The alanin (A) at position 130 in the query is mismatched with the aspartate (D) at position 97 in the subject, as we had seen before.

Answer 116 As the PAM number goes toward infinity, the scores for homologous pairs of amino acids all become 0. The last pair of amino acids with a score > 0 is tryptophan/tryptophan in PAM2000, which means this is the most conserved, or most slowly evolving, amino acid.

```
pamPower -n 1000 pam1.txt  |
pamNormalize -a aa.txt     |
pamLog
```

	A	R	N	D	C	Q	E	G	H	I	L	K	M	F	P	S	T	W	Y	V
A	0	0	0	0	0	0	0	0	0	0	0	0	0	0	0	0	0	-1	0	0
R	0	0	0	0	0	0	0	0	0	0	0	0	0	0	0	0	0	0	0	0
N	0	0	0	0	0	0	0	0	0	0	0	0	0	0	0	0	0	0	0	0
D	0	0	0	0	0	0	0	0	0	0	0	0	0	0	0	0	0	-1	0	0
C	0	0	0	0	3	0	0	0	0	0	0	0	0	0	0	0	0	-1	0	0
Q	0	0	0	0	0	0	0	0	0	0	0	0	0	0	0	0	0	0	0	0
E	0	0	0	0	0	0	0	0	0	0	0	0	0	0	0	0	0	-1	0	0
G	0	0	0	0	0	0	0	0	0	0	0	0	0	0	0	0	0	-1	0	0
H	0	0	0	0	0	0	0	0	0	0	0	0	0	0	0	0	0	0	0	0
I	0	0	0	0	0	0	0	0	0	0	0	0	0	0	0	0	0	0	0	0
L	0	0	0	0	0	0	0	0	0	0	0	0	1	0	0	0	0	0	0	0
K	0	0	0	0	0	0	0	0	0	0	0	0	0	0	0	0	0	0	0	0
M	0	0	0	0	0	0	0	0	0	0	0	0	0	0	0	0	0	0	0	0
F	0	0	0	0	0	0	0	0	0	0	1	0	0	2	0	0	0	1	1	0
P	0	0	0	0	0	0	0	0	0	0	0	0	0	0	0	0	0	0	0	0
S	0	0	0	0	0	0	0	0	0	0	0	0	0	0	0	0	0	0	0	0
T	0	0	0	0	0	0	0	0	0	0	0	0	0	0	0	0	0	0	0	0
W	-1	0	0	-1	-1	0	-1	-1	0	0	0	0	0	1	0	0	0	7	1	-1
Y	0	0	0	0	0	0	0	0	0	0	0	0	0	1	0	0	0	1	2	0
V	0	0	0	0	0	0	0	0	0	0	0	0	0	0	0	0	0	-1	0	0

```
pamPower -n 2000 pam1.txt  |
pamNormalize -a aa.txt     |
pamLog
```

	A	R	N	D	C	Q	E	G	H	I	L	K	M	F	P	S	T	W	Y	V
A	0	0	0	0	0	0	0	0	0	0	0	0	0	0	0	0	0	0	0	0
R	0	0	0	0	0	0	0	0	0	0	0	0	0	0	0	0	0	0	0	0
N	0	0	0	0	0	0	0	0	0	0	0	0	0	0	0	0	0	0	0	0
D	0	0	0	0	0	0	0	0	0	0	0	0	0	0	0	0	0	0	0	0
C	0	0	0	0	0	0	0	0	0	0	0	0	0	0	0	0	0	0	0	0
Q	0	0	0	0	0	0	0	0	0	0	0	0	0	0	0	0	0	0	0	0
E	0	0	0	0	0	0	0	0	0	0	0	0	0	0	0	0	0	0	0	0
G	0	0	0	0	0	0	0	0	0	0	0	0	0	0	0	0	0	0	0	0
H	0	0	0	0	0	0	0	0	0	0	0	0	0	0	0	0	0	0	0	0
I	0	0	0	0	0	0	0	0	0	0	0	0	0	0	0	0	0	0	0	0
L	0	0	0	0	0	0	0	0	0	0	0	0	0	0	0	0	0	0	0	0
K	0	0	0	0	0	0	0	0	0	0	0	0	0	0	0	0	0	0	0	0
M	0	0	0	0	0	0	0	0	0	0	0	0	0	0	0	0	0	0	0	0
F	0	0	0	0	0	0	0	0	0	0	0	0	0	0	0	0	0	0	0	0
P	0	0	0	0	0	0	0	0	0	0	0	0	0	0	0	0	0	0	0	0
S	0	0	0	0	0	0	0	0	0	0	0	0	0	0	0	0	0	0	0	0
T	0	0	0	0	0	0	0	0	0	0	0	0	0	0	0	0	0	0	0	0
W	0	0	0	0	0	0	0	0	0	0	0	0	0	0	0	0	0	2	0	0
Y	0	0	0	0	0	0	0	0	0	0	0	0	0	0	0	0	0	0	0	0
V	0	0	0	0	0	0	0	0	0	0	0	0	0	0	0	0	0	0	0	0

```
pamPower -n 3000 pam1.txt  |
pamNormalize -a aa.txt     |
pamLog
```

	A	R	N	D	C	Q	E	G	H	I	L	K	M	F	P	S	T	W	Y	V
A	0	0	0	0	0	0	0	0	0	0	0	0	0	0	0	0	0	0	0	0
R	0	0	0	0	0	0	0	0	0	0	0	0	0	0	0	0	0	0	0	0
N	0	0	0	0	0	0	0	0	0	0	0	0	0	0	0	0	0	0	0	0
D	0	0	0	0	0	0	0	0	0	0	0	0	0	0	0	0	0	0	0	0
C	0	0	0	0	0	0	0	0	0	0	0	0	0	0	0	0	0	0	0	0
Q	0	0	0	0	0	0	0	0	0	0	0	0	0	0	0	0	0	0	0	0
E	0	0	0	0	0	0	0	0	0	0	0	0	0	0	0	0	0	0	0	0
G	0	0	0	0	0	0	0	0	0	0	0	0	0	0	0	0	0	0	0	0
H	0	0	0	0	0	0	0	0	0	0	0	0	0	0	0	0	0	0	0	0
I	0	0	0	0	0	0	0	0	0	0	0	0	0	0	0	0	0	0	0	0
L	0	0	0	0	0	0	0	0	0	0	0	0	0	0	0	0	0	0	0	0
K	0	0	0	0	0	0	0	0	0	0	0	0	0	0	0	0	0	0	0	0
M	0	0	0	0	0	0	0	0	0	0	0	0	0	0	0	0	0	0	0	0
F	0	0	0	0	0	0	0	0	0	0	0	0	0	0	0	0	0	0	0	0
P	0	0	0	0	0	0	0	0	0	0	0	0	0	0	0	0	0	0	0	0
S	0	0	0	0	0	0	0	0	0	0	0	0	0	0	0	0	0	0	0	0
T	0	0	0	0	0	0	0	0	0	0	0	0	0	0	0	0	0	0	0	0
W	0	0	0	0	0	0	0	0	0	0	0	0	0	0	0	0	0	0	0	0
Y	0	0	0	0	0	0	0	0	0	0	0	0	0	0	0	0	0	0	0	0
V	0	0	0	0	0	0	0	0	0	0	0	0	0	0	0	0	0	0	0	0

Answer 117 With `pam1000sm.txt`, the C-terminal end of the alignment is changed; the rest of the alignment remains the same. However, with `pam2000sm.txt` and `pam3000sm.txt`, the alignment consists almost entirely of mismatches. Homologous positions cannot be distinguished from random matches using these score matrices.

Answer 118 We can type

```
bash pamPower2.sh | gnuplot -p pamPower.gp
```

where `pamPower2.sh` is

```
for f in approx percentDiff
do
    for a in 1 2 5 10 20 50 100 200 500 1000
    do
        echo -n $a ' '
        pamPower -n $a pam1.txt       |
          tail -n +2                  |
          awk -f ${f}.awk
    done
    echo ' '
done
```

and `approx.awk` is essentially the AWK code from Problem 107:

```
{
    c++
    s += $(c+1) * 0.05
}END{
    print (1-s) * 100
}
```

The gnuplot-script `pamPower.gp` is unchanged—which illustrates the usefulness of storing more complex plot commands in a file. The final plot shows that the approximation is very similar to the exact computation:

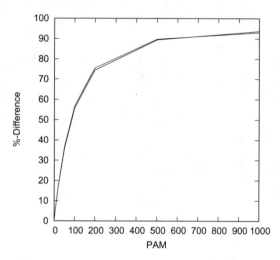

Answer 119 Here is the desired tree:

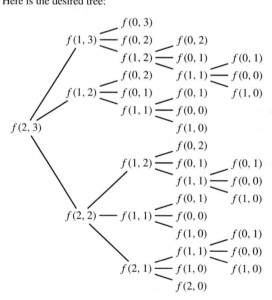

The end result is the number of leaves in that tree, that is nodes without descendants. There are 25 leaves and hence 25 possible global alignments between two sequences of lengths 2 and 3.

Answer 120

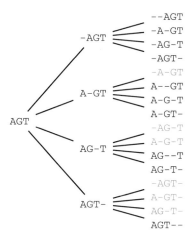

This time we obtain 25 much quicker.

Answer 121 The first thing to realize is that no global alignment of S_1 and S_2 can be shorter than the longer sequence, that is 3, or longer than the sum of their lengths, that is, 5. This means that, say, S_1 can only occur with 0, 1, or 2 gaps. We can write down the longer sequence with 0 gaps and then construct all sequences with 1 gap, and again from these the sequences with 2 gaps. For the latter, we need to watch out for repeats, which are shown in gray:

```
                                                      --AGT
                                                      -A-GT
                                          -AGT        -AG-T
                                                      -AGT-
                                                      -A-GT
                                          A-GT        A--GT
                                                      A-G-T
                                                      A-GT-
                            AGT                        -AG-T
                                                      A-G-T
                                          AG-T         AG--T
                                                      AG-T-
                                                      -AGT-
                                                      A-GT-
                                          AGT-        A-GT-
                                                      AGT--
```

From these, we can construct all 25 alignments:

AGT	AGT	AGT	-AGT	-AGT
AC-	A-C	-AC	AC--	A-C-
-AGT	A-GT	A-GT	A-GT	AG-T
A--C	-AC-	-A-C	AC--	-AC-
AG-T	AG-T	AGT-	AGT-	AGT-
A-C-	--AC	--AC	-A-C	A-C
--AGT	-A-GT	-AG-T	-AGT-	A--GT
AC---	A-C--	A--C-	A---C	-AC-
A-G-T	A-GT-	AG--T	AG-T-	AGT--
-A-C-	-A--C	--AC-	--A-C	---AC

Answer 122 The command

```
numAl -m 106 -n 106
```

tells us there are 7.8×10^{79} possible alignments. This is also approximately the number of atoms in the observable universe spcitewik:obs.

Answer 123 Run

```
bash numAl.sh              |
awk '{print ++c, $3}'  |
gnuplot -p -e 'set xl "Length (bp)"; set yl "Number of
    Alignments"; set log y; p "< cat" t "" w l'
```

where `numAl.sh` is

```
for a in $(seq 106)
do
    numAl -m $a -n $a
done
```

to get

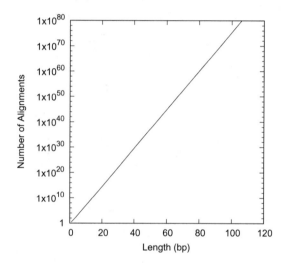

Answer 124 Run

```
bash numAl2.sh
```

where `numAl2.sh` is

```
for a in $(seq 106)
do
    numAl -t -m $a -n $a
done
```

to observe that the computations quickly become very slow. We gave up at length 14.

Answer 125 To make subsequent manipulations more convenient, first write the run times
to a file

```
for a in $(seq 14); do numAl -t -m $a -n $a; done > times.
    txt
```

Then filter out the times greater than zero and plot them:

```
cat times.txt                    |
awk '{if($6>0)print ++c, $6}'    |
gnuplot -p -e 'set xl "Length (bp)"; set yl "Time (s)"; set
     log y; p "< cat" t "" w l'
```

to get

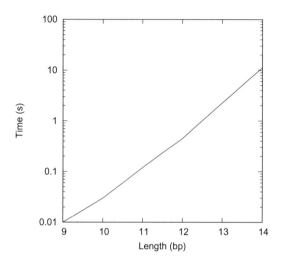

Answer 126 Linear functions are described by

$$f(x) = ax + b.$$

The slope of the straight line, a, is found by considering, for example, the run times for sequences lengths 13 and 14, which were 2.25 s and 11.33 s in our case. Take logarithms

```
awk 'BEGIN{print log(2.25)/log(10)}'
0.352183
awk 'BEGIN{print log(11.33)/log(10)}'
1.05423
```

to get $a = 1.05 - 0.35 = 0.70$ and $b = 1.05 - 14 \times 0.7 = -8.75$. So our run time is

$$10^{0.7 \times 106 - 8.75} = 10^{65.45}$$

seconds or

$$\frac{10^{65.45}}{60 \times 60 \times 24 \times 365.25} \approx 8.9 \times 10^{57}$$

years. This dwarfs the 10^{10} years the universe has existed. Choosing the appropriate algorithm can make an enormous difference.

Answer 127

Here is our dot plot:

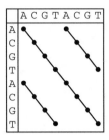

The matches off the main diagonal indicate duplications.

Answer 128 The fist two lines of dmAdhAdhdup.fasta are

```
head -n 2 dmAdhAdhdup.fasta
>DMADH X78384.1 D.melanogaster Adh and Adh-dup genes.
TGTATTTTCCAATTAGGTGATAGAACTTGTGTGCACACACACATATAGTTCTATATCAAC
```

So dmAdhAdhdup.fasta contains the genes *Adh* and its duplication *Adh-dup* from *D. melanogaster*. The first two lines of dgAdhAdhdup.fasta are

```
head -n 2 dgAdhAdhdup.fasta
>DGADHDUP X60113.1 D.guanche Adh and Adh-dup genes for
    alcohol dehydrogenase.
TCTAGATTGCATCACTCGTGCCGCCCTACGTTGTGAAGCACCACGCCCTGGACCCCGTTT
```

Correspondingly, dgAdhAdhdup.fasta contains the *D. guanche* versions of *Adh* and *Adh-dup*. Now count the nucleotides:

```
cchar dmAdhAdhdup.fasta
# Total number of input characters: 4761
# Char   Count    Fraction
A        1417     0.297627
C        1007     0.211510
G        989      0.207729
T        1348     0.283134
```

and

```
cchar dgAdhAdhdup.fasta
# Total number of input characters: 4433
# Char   Count    Fraction
A        1279     0.288518
C        998      0.225130
G        936      0.211144
T        1220     0.275209
```

In summary,

File	Genes	Organism	Length (bp)
dmAdhAdhdup.fasta	*Adh &Adh-dup*	*D. melanogaster*	4761
dgAdhAdhdup.fasta	*Adh & Adh-dup*	*D. guanche*	4433

Answer 129 The two sequences are 4761 and 4433 bp long, and the corresponding dot plot would therefore contain $4761 \times 4433 = 21, 105, 513$ cells. This is quite a large number. We therefore need an approach that is more efficient than checking each cell and placing a dot for every match.

Answer 130 Run

```
#len|strId:pos_1|...|strId:pos_n|seq
37|f2:3292|f1:3287|AGCAAGGTTCTCATGACCAAGAATATAGCGGTGAGTG
```

The longest repeat has length 37.

Answer 131 When running

```
cat *.fasta | randomizeSeq | repeater
```

a couple of times, we commonly find repeats with lengths 11 to 14. Since these lengths are not even close to the naturally occurring longest repeat of 37, it is extremely unlikely that this repeat has occurred by chance.

Answer 132 The raw result is

```
cat dmAdhAdhdup.fasta dgAdhAdhdup.fasta |
repeater -m 12                          |
head -n 3
#len|strId:pos_1|...|strId:pos_n|seq
13|f2:3096|f1:3129|AAAATAGATAAAT
12|f2:3371|f1:3389|AAACTAATTAAG
```

This is converted by `dotPlotFilter.awk` to

```
cat dmAdhAdhdup.fasta dgAdhAdhdup.fasta |
repeater -m 12                          |
head -n 3                               |
awk -f dotPlotFilter.awk
3129    3096
3141    3108

3389    3371
3400    3382
```

For a given repeat, `dotPlotFilter.awk` prints the x- and y-coordinates of the start and end of the repeat separated by blank lines.

Answer 133 Commands like

```
cat dmAdhAdhdup.fasta dgAdhAdhdup.fasta |
repeater -m 12                          |
awk -f dotPlotFilter.awk                |
gnuplot -p -e 'set xl "D. melanogaster"; set yl "D. guanche
     "; p "< cat" t "" w l'
```

give results like

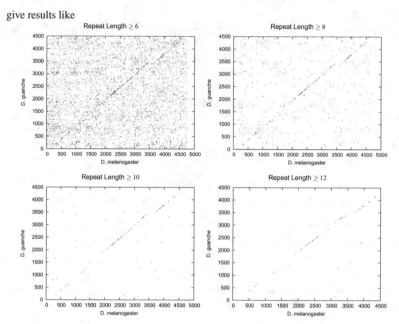

The repeat length determines the density of the plot.

Answer 134 Since the paralogs are contained in both species, the duplication must predate the speciation event. Accordingly, the orthologs have a more recent last common ancestor than the paralogs.

Answer 135 A cartoon phylogeny would look like this:

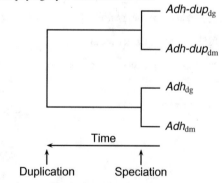

The branch lengths are unknown, but the branch order *is* known, and hence the relative order of gene divergence due to duplication and speciation.

Answer 136 The homology between the orthologous pairs of genes is clearly visible along the main diagonal of the plot. However, there are no significant off-diagonal repeats, and hence the paralogous relationships between the duplicated genes is so ancient as to have become invisible in a dot plot.

Answer 137 The insertion, marked by an arrow below, causes a shift in the main diagonal of homology:

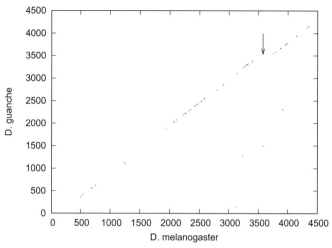

There was an insertion into the *D. melanogaster* sequence.

Answer 138 The coordinates for *D. melanogaster* are found by typing

```
grep CDS dmAdhAdhdup.gb
```

and for *D. guanche* by entering

```
grep CDS dgAdhAdhdup.gb
```

to get

Organism	*Adh*	*Adh-dup*
D. melanogaster	2021..2119,2185..2589,2660..2926	3226..3321,3748..4152,4204..4521
D. guanche	1984..2076,2145..2549,2613..2879	3221..3316,3540..3944,4007..4345

Answer 139 There are many ways of converting the CDS coordinates recorded in the genbank file to a set of intervals. Our version is

```
grep CDS dmAdhAdhdup.gb |
sed -f cds.sed          |
awk -f break.awk > cdsDm.txt
```

where `cds.sed` is

```
s/.*(//       # remove everything up to opening bracket
s/)//         # remove closing bracket
s/\.\./ /g    # substitute blanks for pairs of dots
s/,/ /g       # substitute blanks for commas
```

and `break.awk` is

```
{
    printf "%s\t%s\n%s\t%s\n%s\t%s\n", $1, $2, $3, $4, $5, $6
}
```

Next, write the dot plot data to file:

```
cat dmAdhAdhdup.fasta dgAdhAdhdup.fasta |
repeater -m 12                          |
awk -f dotPlotFilter.awk > dotPlot.dat
```

Append the exon coordinates:

```
awk -f boxesX.awk cdsDm.txt >> dotPlot.dat
```

where `boxesX.awk` is

```
BEGIN{
    h = 150 # height
}{
    s = $1  # start
    e = $2  # end
    printf("%d\t%d\n%d\t%d\n%d\t%d\n%d\t%d\n", s, 0, s, h, e
        , h, e, 0)
}
```

and `cdsDm.txt` contains the CDS coordinates of *D. melanogaster*:

```
2021 2119
2185 2589
2660 2926
3226 3321
3748 4152
4204 4521
```

Now plot the result:

```
cat dotPlot.dat |
gnuplot -p -e 'set xl "D. melanogaster"; set yl "D. guanche
    "; p "< cat" t "" w l'
```

to get

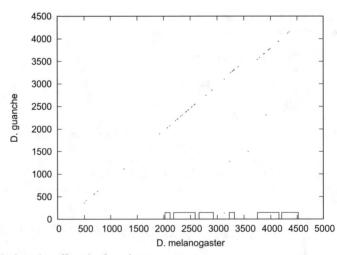

It seems the insertion affected only an intron.

Answer 140 Draw the plot as

```
gnuplot -p cdsDm.gp
```

where `cdsDm.gp` is

```
set xlabel "D. melanogaster"
set ylabel "D. guanche"
set arrow from 3400,0 to 3400,4500 nohead
set arrow from 3740,0 to 3740,4500 nohead
plot "dotPlot.dat" title "" with lines
```

We plotted directly from the data file `dotPlot.dat`; alternatively, it could have been read from the standard input. However, in a script it is usually more convenient to read from a file containing the input data. By drawing vertical lines along the gap borders, we can clearly see that the insertion affected an intron:

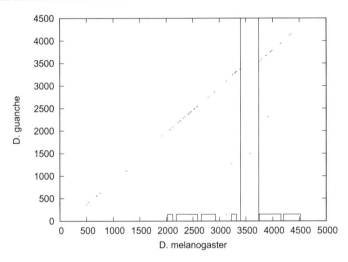

Answer 141 Generate the exon coordinates for *D. guanche*:

```
grep CDS dgAdhAdhdup.gb |
sed -f cds.sed           |
awk -f break.awk > cdsDg.txt
```

Add an empty line to the data file

```
echo >> dotPlot.dat
```

Add the exons for *D. guanche*:

```
awk -f boxesY.awk cdsDg.txt >> dotPlot.dat
```

where `boxesY.awk` is

```
BEGIN{
    h = 150 # height
}{
    s = $1  # start
    e = $2  # end
    printf("%d\t%d\n%d\t%d\n%d\t%d\n%d\t%d\n", 0, s, h, s, h,
        e, 0, e)
}
```

Now plot the data

```
gnuplot -p adhCds.gp
```

where the `gnuplot` script `adhCds.gp` is

```
set termoption dash
reset
set xlabel "D. melanogaster"
set ylabel "D. guanche"
# D. m. CDS
set arrow from 2021,0 to 2021,4500 nohead lt 2
set arrow from 2926,0 to 2926,4500 nohead
set arrow from 3226,0 to 3226,4500 nohead
set arrow from 4521,0 to 4521,4500 nohead
# D. g. CDS
set arrow from 0,1984 to 5000,1984 nohead
set arrow from 0,2879 to 5000,2879 nohead
set arrow from 0,3221 to 5000,3221 nohead
set arrow from 0,4345 to 5000,4345 nohead
# plot
plot "dotPlot.dat" title "" with lines
```

to get the CDS of *Adh* and *Adh-dup* for *D. melanogaster* and *D. guanche*:

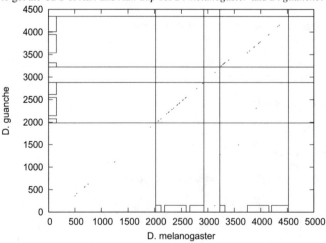

Answer 142 Here is the dot plot:

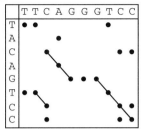

There are three possible global alignments with the maximum number of matches:

```
TTCAGGGTCC        TTCAGGGTCC        TTCAGGGTCC
TACA--GTCC        TACA-G-TCC        TACAG--TCC
```

The last column of these alignments corresponds the bottom right corner of the dot plot matrix.

Answer 143 An alignment gap corresponds to a shift in a match diagonal.

Answer 144 $F(2, 1)$ refers to $S_1[1..1] = $ A and $S_2[1..2] = $ AC.

Answer 145

	–	A	C	T
	0	1	2	3
– 0	0 ←	−1 ←	−2 ←	−3
A 1				
C 2				

Answer 146

	–	A	C	T
	0	1	2	3
– 0	0	← −1	← −2	← −3
A 1	↑ −1			
C 2	↑ −2			

Answer 147

	–	A	C	T
	0	1	2	3
– 0	0	← −1	← −2	← −3
A 1	↑ −1	↖ 1	← 0	← −1
C 2	↑ −2	↑ 0	↖ 2	← 1

Answer 148 The traceback returns

$$ACT$$
$$AC-$$

This has score 1, as expected from the entry in the lower right-hand cell of the alignment matrix.

Answer 149 In a global alignment, the bottom right-hand cell of the alignment matrix contains the score of the optimal alignment between S_1 and S_2; in our case, this is 4.

Answer 150 There are three cooptimal alignments implied by the global alignment matrix:

```
TTCAGGGTCC
| ||  ||||
TACA--GTCC
```

```
TTCAGGGTCC
| || | |||
TACA-G-TCC
```

```
TTCAGGGTCC
| |||  |||
TACAG--TCC
```

all of which have score 4, the entry in the cell from where the traceback started.

Answer 151 Construct directory and change into it:

```
mkdir OptimalAlignment
cd OptimalAlignment
```

Run `gal` with the desired score scheme:

```
gal -i seq1.fasta -j seq2.fasta -O 0 -E -1 -I -1
```

```
Query:              >seq1
   Length:          10
Subject:            >seq2
   Length:          8
Score:              4.0
Strand:             Plus / Plus


Query: 1 TTCAGGGTCC 10
         | ||  ||||
Sbjct: 1 TACA--GTCC 8
```

This can be changed; for example, if the mismatch score is ≤ -3 and all other parameters are left unchanged, the optimal alignment becomes

```
TT-CAGGGTCC
 | ||  ||||
-TACA--GTCC
```

Answer 152 Again, the answer is yes, but only for gap score schemes lacking biological plausibility: For example, we might reward gaps by setting the gap extension parameter to 1. In other words, the score increases rather than decreases when gaps are inserted. The alignment then becomes

```
gal -E 1 -i s1.fasta -j s2.fasta
TTCAGGGT-----CC
        |     ||
-------TACAGTCC
```

Answer 153 The *Adh-dup* from *D. guanche* should contain a large gap, which corresponds to an insertion in *D. melanogaster*.

Answer 154 Enter

```
gal -i dmAdhAdhdup.fasta -j dgAdhAdhdup.fasta | less
```

and find that between positions 3426 and 3531 in *D. guanche* there are large gaps. Now look up the CDS coordinates for *D. guanche*

```
grep CDS dgAdhAdhdup.gb
CDS join(1984..2076,2145..2549,2613..2879)
CDS join(3221..3316,3540..3944,4007..4345)
```

The first CDS refers to *Adh*, the second to *Adh-dup*. The intron between the first and the second exon of *Adh-dup* has coordinates 3317–3539, which corresponds to the gapped region. Therefore, the gap is located in the first intron of the *D. guanche Adh-dup*.

Answer 155 First initialize, the fill in the local alignment matrix:

Initialize						
	-	T	A	C	G	T
	0	1	2	3	4	5
- 0	0	0	0	0	0	0
G 1	0					
A 2	0					
C 3	0					
G 4	0					
A 5	0					

Fill in						
	-	T	A	C	G	T
	0	1	2	3	4	5
- 0	0	0	0	0	0	0
G 1	0	0	0	0	↖1	0
A 2	0	0	↖1	0	0	0
C 3	0	0	0	↖2	←1	0
G 4	0	0	0	↑1	↖3	←2
A 5	0	0	↖1	0	↑2	↖2

Answer 156 The traceback looks like this:

	-	T	A	C	G	T
	0	1	2	3	4	5
- 0	0	0	0	0	0	0
G 1	0	0	0	0	↖1	0
A 2	0	0	↖1	0	0	0
C 3	0	0	0	↖2	←1	0
G 4	0	0	0	↑1	↖3	←2
A 5	0	0	↖1	0	↑2	↖2

The resulting alignment is

ACG
ACG

Its score is 3, the entry in the cell from which the traceback started.

Answer 157 The command

```
lal -i dmAdhAdhdup.fasta -j dgAdhAdhdup.fasta
```

returns

```
Query:                >DMADH X78384.1 D.MELANOGASTER ADH AND ADH
                      -DUP GENES.
  Length:             4761
  Hit:                2182 - 2594
Subject:              >DGADHDUP X60113.1 D.guanche Adh and Adh-
                      dup genes for alcohol dehydrogenase.
  Length:             4433
  Hit:                2142 - 2554
Score:                217.0
Strand:               Plus / Plus
```

```
Query: 2182 TAGAACCTGGTGATCCTCGACCGCATTGAGAACCCGGCTGCCATTGCCGAGCTGAAGGCA 2241
            |||||||||||| ||||| || |||||||| || || |||||||||||||| ||||||||
Sbjct: 2142 TAGAACCTGGTCATCCTGGATCGCATTGACAATCCAGCTGCCATTGCCGAACTGAAGGCA 2201


Query: 2242 ATCAATCCAAAGGTGACCGTCACCTTCTACCCCTATGATGTGACCGTGCCCATTGCCGAG 2301
            ||||||| |||||||||||||||||||||||| |||||||||| || || | || |||
Sbjct: 2202 GTCAATCCCAAGGTGACCGTCACCTTCTACCCTTATGATGTGACTGTACCTGTCGCAGAG 2261


Query: 2302 ACCACCAAGCTGCTGAAGACCATCTTCGCCCAGCTGAAGACCGTCGATGTCCTGATCAAC 2361
            |||||||| || |||||||||||| |||||| | |||||| |||||||||||||| |||
Sbjct: 2262 ACCACCAAACTCCTGAAGACCATCTTTGCCCAGATCAAGACCATCGATGTCCTGATAAAC 2321


Query: 2362 GGAGCTGGTATCCTGGACGATCACCAGATCGAGCGCACCATTGCCGTCAACTACACTGGC 2421
            || ||||| ||||| |||||||| ||||| ||||| || |||||||| ||||||||||||
Sbjct: 2322 GGTGCTGGCATCCTCGACGATCATCAGATTGAGCGTACTATTGCCGTTAACTACACTGGC 2381


Query: 2422 CTGGTCAACACCACGACGGCCATTCTGGACTTCTGGGACAAGCGCAAGGGCGGTCCCGGT 2481
            ||||||||||||||| || |||||||||| ||||||||||||||||||||||||| || |||
Sbjct: 2382 CTGGTCAACACCACCACAGCCATTCTGGATTTCTGGGACAAGCGCAAGGGCGGCCCAGGT 2441


Query: 2482 GGTATCATCTGCAACATTGGATCCGTCACTGGATTCAATGCCATCTACCAGGTGCCCGTC 2541
            || ||||| ||||||||||| ||||| || || || |||||||||||||||||||||||
Sbjct: 2442 GGCATCATTTGCAACATTGGCTCCGTTACCGGTTTTAATGCCATCTACCAGGTGCCCGTT 2501


Query: 2542 TACTCCGGCACCAAGGCCGCCGTGGTCAACTTCACCAGCTCCCTGGCGGTAAG 2594
            ||||| |||| |||||| || ||||| ||||||||||||||||||||||||||
Sbjct: 2502 TACTCTGGCAGCAAGGCGGCCGGTGGTAAACTTCACCAGCTCCCTGGCGGTAAG 2554
```

Hence, the coordinates of the best local alignment are as follows:

Organism	Local alignment
D. melanogaster	2182–2594
D. guanche	2142–2554

As before, we look up the coordinates of the CDS for *Adh* and *Adh-dup* in *D. melanogaster*

```
grep CDS dmAdhAdhdup.gb
    CDS             join(2021..2119,2185..2589,2660..2926)
    CDS             join(3226..3321,3748..4152,4204..4521)
```

and similarly for *D. guanche*

```
grep CDS dgAdhAdhdup.gb
    CDS                  join(1984..2076,2145..2549,2613..2879)
    CDS                  join(3221..3316,3540..3944,4007..4345)
```

to find that the best alignment corresponds to the second exon of *Adh*.

Answer 158 Compute a single alignment:

```
time lal -i dmAdhAdhdup.fasta -j dgAdhAdhdup.fasta
```

This takes approximately 2.8 s, while computing two optimal local alignments

```
time lal -n 2 -i dmAdhAdhdup.fasta -j dgAdhAdhdup.fasta
```

takes approximately 22.5 s, eight times longer. The reason for this is that the top local alignment is surrounded by many cells with high scores. But the paths starting from these cells all intersect the optimal local alignment. It is therefore hard for the algorithm to find the starting point of the next best distinct alignment.

Answer 159 The coordinates of the second best alignment are as follows:

Organism	Local alignment
D. melanogaster	3829–4162
D. guanche	3621–3954

As before, look up the CDS coordinates

```
grep CDS *.gb
```

to find that the second best alignment corresponds to exon 2 of *Adh-dup*.

Answer 160 The score of the alignment is

```
gal -i dmAdhCds.fasta -j dmAdhdupCds.fasta |
grep Score
Score:                  -1631.0
```

Answer 161 Compute the scores from random alignments

```
randomizeSeq -n 1000 dmAdhCds.fasta    |
gal -i dmAdhdupCds.fasta                |
grep Score                             |
awk '{print $2}' > scores.dat
```

Plot the scores

```
histogram scores.dat | gnuplot -p scores.gp
```

where scores.gp is

```
set arrow from -1631,0.01 to -1631,0 # observed
set xl "Score"
set yl "Frequency"
plot[-1800:-1600][] "< cat" title "" with lines
```

to get

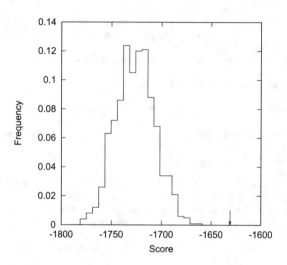

It is highly unlikely that the observed score (arrow) occurs between shuffled, nonhomologous versions of the input sequences.

Answer 162 First, remind yourself of the length of dmAdhAdhdup.fasta using cchar:

```
cchar dmAdhAdhdup.fasta
# Total number of input characters: 4761
```

Cut out the first and last kb:

```
cutSeq -r 1-1000     dmAdhAdhdup.fasta > f1.fasta
cutSeq -r 3762-4761 dmAdhAdhdup.fasta > f2.fasta
```

Compute the original score

```
gal -i f1.fasta -j f2.fasta | grep Score
Score:              -1385.0
```

Compute the random scores:

```
randomizeSeq -n 1000 f2.fasta |
gal -i f1.fasta               |
grep Score                    |
awk '{print $2}' > scoresR.dat
```

Plot the random scores

```
histogram scoresR.dat | gnuplot -p scoresR.gp
```

where scoresR.gp is

```
set arrow from -1385,0.01 to -1385,0 # observed
set xl "Score"
set yl "Frequency"
plot[][] "< cat" title "" with lines
```

to get

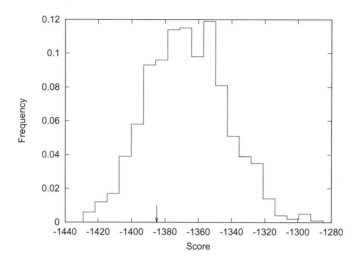

As expected, the score between the two arbitrary DNA segments (arrow) is not different from random. In other words, only scores to the right of the distribution of random scores are indicative of homology.

Answer 163 The substitution rates K_1 and K_2 are between orthologs, and thus refer to the split between *D. melanogaster* and *D. guanche*. The other four are between paralogs and refer to the *Adh* duplication.

Answer 164 Extract exon 2 from *Adh* and *Adh-dup* of *D. melanogaster*:

```
cutSeq -r 2185-2589 dmAdhAdhdup.fasta > dmAdhE2.fasta
cutSeq -r 3748-4152 dmAdhAdhdup.fasta > dmAdhdupE2.fasta
```

Repeat for *D. guanche*:

```
cutSeq -r 2145-2549 dgAdhAdhdup.fasta > dgAdhE2.fasta
cutSeq -r 3540-3944 dgAdhAdhdup.fasta > dgAdhdupE2.fasta
```

Answer 165

```
gal -i dmAdhE2.fasta -j dgAdhE2.fasta |
sed -n '/|/p'                          |
sed 's/ //g'                           |
awk '{s+=length($1)}END{print s}'
356
```

Answer 166 Exon 2 is 405 bp long and there were 356 matches. So the number of mismatches per site is

$$\pi = (405 - 356)/405 \approx 0.12.$$

Answer 167 Convert 0.12 mismatches per site to the number of substitutions per site:

$$K_1 = -\frac{3}{4} \log\left(1 - \frac{4}{3} \times 0.12\right) \approx 0.13.$$

Answer 168 Compute the number of matches

```
gal -i dmAdhdupE2.fasta -j dgAdhdupE2.fasta |
sed -n '/|/p'                                |
sed 's/ //g' | awk '{s+=length($1)}END{print s}'
324
```

So the number of mismatches per site is

$$\pi_2 = (405 - 324)/405 \approx 0.2,$$

and

$$K_2 = -\frac{3}{4}\log\left(1 - \frac{4}{3}\pi_2\right) \approx 0.23.$$

Answer 169 To compute M_3, we use the same commands as before:

```
gal -i dmAdhE2.fasta -j dmAdhdupE2.fasta |
sed -n '/|/p'                            |
sed 's/ //g'                             |
awk '{s+=length($1)}END{print s}'
```

We get

Comparison	Matches
$M_3 : Adh_{dm} / Adh\text{-}dup_{dm}$	228
$M_4 : Adh\text{-}dup_{dg} / Adh_{dg}$	224
$M_5 : Adh_{dm} / Adh\text{-}dup_{dg}$	235
$M_6 : Adh\text{-}dup_{dm} / Adh_{dg}$	226

That is, on average there are $M_{dup} = (228 + 224 + 235 + 226)/4 = 228.25$ matches. The number of mismatches per site is thus $\pi_{dup} = (405-228.25)/405 \approx 0.436$ and $K_{dup} \approx 0.65$.

Answer 170 The average between K_1 and K_2 is $(0.13+0.23)/2 = 0.18$. A rough estimate of the age of duplication is

$$0.65/0.18 \times 32 \approx 116$$

million years. When drawn as a tree, these divergence times look like this:

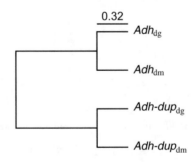

Answer 171 Take care of the preliminaries

```
mkdir KeywordTrees
cd KeywordTrees
```

and write the program `split.awk`:

```
BEGIN{
    n = split(t, ta, "")
    for(i=1; i<=n; i++)
        print ta[i]
}
```

Answer 172 Run

```
awk -v t=CACAGACACAT -v p=ACA -f naive.awk
```

where `naive.awk` is

```
BEGIN{
    n = split(t, ta, "")
    m = split(p, pa, "")
    for(i=1; i<=n-m; i++){
        for(j=1; j<=m; j++){
            if(ta[i+j-1] != pa[j])
                break
        }
        if(j == m + 1)
            print i
    }
}
```

Answer 173 Run

```
echo -e '>Seq\nACGTCG' |
awk -f naive2.awk -v file=mgGenome.fasta
122599
```

where `naive2.awk` is

```
BEGIN{
    cmd = "tail -n +2 " file
    while(cmd | getline)
        t = t $1
}
!/^>/{
    p = p $1
}
END{
    n = split(t, ta, "")
    m = split(p, pa, "")
    for(i=1; i<=n-m; i++){
        for(j=1; j<=m; j++){
            if(ta[i+j-1] != pa[j])
                break
        }
        if(j == m + 1)
            print i
    }
}
```

Answer 174 Reverse complement the sequence:

```
revComp mgGenome.fasta > mgGenomeR.fasta
```

and run

```
echo -e '>Seq\nCGGCCT' |
awk -f naive2.awk -v file=mgGenomeR.fasta
270306
```

to find one copy of the motif on the reverse strand.

Answer 175 Run

```
awk -f monoNuc.awk -v n=100 | fold
```

where monoNuc.awk is

```
BEGIN{
    print ">Mononuc"
    for(i=0; i<n; i++)
        printf("A")
    printf("\n")
}
```

Answer 176 Generate the text files

```
awk -f monoNuc.awk -v n=1000000 | fold > 1mb.fasta
awk -f monoNuc.awk -v n=2000000 | fold > 2mb.fasta
```

Measure the run times

```
awk -f monoNuc.awk -v n=10 |
  time awk -f naive2.awk -v file=1mb.fasta | tail
awk -f monoNuc.awk -v n=20 |
  time awk -f naive2.awk -v file=2mb.fasta | tail
```

where we got 2.65 s and 9.56 s, respectively. This roughly 3.6-fold increase in run time is somewhat smaller than the expected $2 \times 2 = 4$-fold increase.

Answer 177 The command

```
time naiveMatcher -p AAAAAAAAAAAAAAAAAAAA 2mb.fasta | tail
```

takes 0.282 s, compared to 9.56 s for the AWK script. This is a 34-fold speedup obtained just by switching from AWK to C.

Answer 178 Run

```
bash runNaive.sh > runNaive.dat
```

where runNaive.sh is

```
for a in 10 20 50 100 200 500 1000 2000 5000 10000
do
    echo -n ${a} ' '
    awk -f monoNuc.awk -v n=${a} |
        fold > pattern.fasta
```

```
        /usr/bin/time -p naiveMatcher -P pattern.fasta 2mb.fasta
          2>&1 |
          grep real |
          sed 's/real //'
done
```

and plot

```
gnuplot -p plot1.gp
```

where `plot1.gp` contains

```
set xlabel "Pattern Length (kb)"
set ylabel "Time (s)"
plot "runNaive.dat" using ($1/1000):2 title "" with lines
```

to get

In the artificial situation of an alphabet consisting of just a single character, the run time of
`naiveMatcher` is linear in the pattern length.

Answer 179

Answer 180

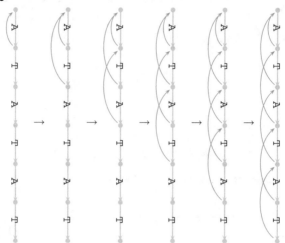

Answer 181 The command

```
/usr/bin/time -p keywordMatcher -p AAAAAAAAAAAAAAAAAAAA 2mb.
    fasta 2>&1 |
grep real
```

runs in 0.40 s. This is actually slower than the naïve version, which took 0.26 s:

```
/usr/bin/time -p naiveMatcher -p AAAAAAAAAAAAAAAAAAAA 2mb.
    fasta 2>&1 |
grep real
```

Answer 182 Execute

```
bash runKeyword.sh > runKeyword.dat
```

where runKeyword.sh is almost identical to runNaive.sh, only the line

```
/usr/bin/time -p naiveMatcher -P pattern.fasta 2mb.fasta 2>&1 |
```

is replaced by

```
/usr/bin/time -p keywordMatcher -f pattern.fasta 2mb.fasta 2>&1|
```

Plot runKeyword.dat together with runNaive.dat using

```
gnuplot -p plot2.gp
```

where plot2.gp is

```
set xlabel "Pattern Length (kb)"
set ylabel "Time (s)"
plot "runNaive.dat" using ($1/1000):2 title "naiveMatcher"
    with lines,\
"runKeyword.dat" using ($1/1000):2 title "keywordMatcher"
    with lines
```

to get

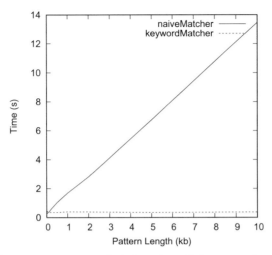

The run time of `keywordMatcher` is independent of the pattern length, while that of `naiveMatcher` is proportional to the pattern length. Notice that this observation only applies to our extreme case, where pattern and text consist of a single type of nucleotide.

Answer 183 We first measure the run times of `naiveMatcher`:

```
bash runNaiveLen.sh > runNaiveLen.dat
```

where `runNaiveLen.sh` is

```
for a in 1 2 5 10 20 50 100
do
    echo -n ${a} ' '
    ranseq -l ${a}000000 > tmp.fasta
    /usr/bin/time -p naiveMatcher -p TTTAACCTCCGGCGGAGTTT
        tmp.fasta 2>&1 |
        grep real |
        sed 's/real //'
done
```

Likewise, we execute

```
bash runKeywordLen.sh > runKeywordLen.dat
```

where `runKeywordLen.sh` is

```
for a in 1 2 5 10 20 50 100
do
    echo -n ${a} ' '
    ranseq -l ${a}000000 > tmp.fasta
    /usr/bin/time -p keywordMatcher -p TTTAACCTCCGGCGGAGTTT
        tmp.fasta 2>&1 |
        grep real |
        sed 's/real //'
done
```

The two result files are plotted

```
gnuplot -p plot3.gp
```

where `plot3.gp` is

```
set xlabel "Text Length (Mb)"
set ylabel "Time (s)"
set key top center
plot "runNaiveLen.dat" title "naiveMatcher" with lines,\
"runKeywordLen.dat" title "keywordMatcher" with lines
```

to get

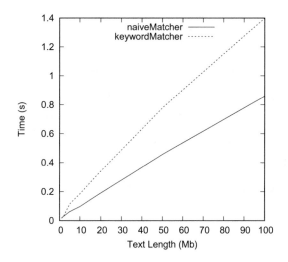

With random sequences, where on average only a small portion of the prefix is checked before a mismatch is found, the simple naïve method beats the more complex method based on pattern preprocessing.

Answer 184 Here is the keyword tree with all failure links included:

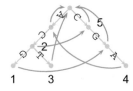

Answer 185 You could enter, for example,

```
keywordMatcher -t kt.tex -p 'ACG|AC|ACT|CGA|C' 2mb.fasta >
   /dev/null
```

followed by

```
latex ktWrapper.tex; dvips ktWrapper.dvi
```

and

```
gv ktWrapper.ps &
```

to get

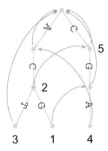

This is not quite as nicely laid out as the manually generated keyword tree, but it does the job. Deleting \date{} results in an automatically generated date produced by \maketitle.

Answer 186 Walking into the tree recovers a match to P_2 = AC, followed by a match to P_1 = ACG. Then, follow the mismatch link twice to find P_5 = C. However, the first occurrence of C is missed that way. To prevent this, the matches at a particular point in the tree are not restricted to the pattern that may end at that point, but to any pattern that can be reached via the failure links. In practice, the preprocessing of a keyword tree includes a step where for every node the failure links are followed, resulting in an output set consisting of the node labels encountered along the way. Each output set may contain zero, one, or several patterns.

Answer 187 For most nodes of our keyword tree, the output set simply consists of the node label; the exception is node 2, where the output set now also includes P_5:

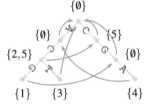

Answer 188

$$
\begin{array}{ll}
S[1..] & \text{ACCCG} \\
S[2..] & \text{CCCG} \\
S[3..] & \text{CCG} \\
S[4..] & \text{CG} \\
S[5..] & \text{G} \\
\end{array}
$$

Answer 189 Here is our first suffix tree:

Answer 190 Start at the root and walk into the tree until CC has been matched. Then, look up the leaf labels of the subtree rooted on the node connected to the edge on which the search ended. These labels indicate the starting positions of CC in S, positions 3 and 2 in our example.

Answer 191 The last suffix, $S[5..] = $ A, does not generate a mismatch and hence no leaf in the suffix tree. This always happens when a suffix is a prefix of another suffix, like in our example where $S[5..] = $ A is a prefix of the suffix $S[1..] = $ ACCCA. To guarantee that each suffix generates a mismatch and hence a leaf when it is threaded into the tree, a so-called sentinel character is added at the end of S: $S = $ ACCCA$. This sentinel character, denoted by $, is not a nucleotide and hence cannot occur anywhere in the sequence.

Answer 192

1 2 3 4 5 6
A C C C A $

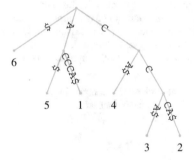

Answer 193 The string depths are marked in red:

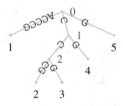

Answer 194 Look for the node with the greatest string depth. Its path label is the longest repeat in S, which is CC in our case.

Answer 195 Assuming your current directory is BiProblems, make the new directory, change into it, and copy the genome file:

```
mkdir SuffixTrees
cd SuffixTrees
cp ../Data/mgGenome.fasta .
```

Then execute

```
repeater -i mgGenome.fasta
#len|strId:pos_1|...|strId:pos_n|seq
243|389491|390403|TTTTTCAGCAGTTGGTTG...
```

to find that the genome of *M. genitalium* contains a repeat of 243 bp that occurs at positions 389,491 and 390,403.

Answer 196

```
# Total number of input characters: 580076
# Char  Count    Fraction
A       200544   0.345720
C       91515    0.157764
G       92306    0.159127
T       195711   0.337389
```

The probability of drawing AA is $0.35^2 \approx 0.12$.

Answer 197

```
cchar mgGenome.fasta   |
sed '/^#/d'            |
awk '{s+=$3^2}END{printf "P_m = %f\n", s}'
P_m = 0.283565
```

Answer 198 We need to solve

$$1 = P_m^l \times L^2$$

for l. By rearranging and taking logarithms, we get

$$l = \frac{\log(1/L^2)}{\log(P_m)}.$$

By substituting $P_m = 0.284$ and $L = 580,076$, we find $l \approx 21.1$. This is much shorter than the observed longest repeat of length 243.

Answer 199 Our basic computation is

```
randomizeSeq mgGenome.fasta | repeater
```

which returns values like 23, 19, 21, 20, 19, 20, 21, 19, and so on. From 100 iterations, we found an average maximum length of 20.1, which is close to the expected 21.1, but still significantly smaller. One reason for this might be that in our model all starting points in the matrix are independent of each other, which is a simplification.

Answer 200 The suffix tree for AAAA looks like this:

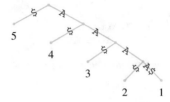

During its construction, each suffix needed to be threaded from its beginning to its end into the intermediate tree. This means that for a sequence of length n,

$$n - 1 + n - 2... + 1 = n(n - 1)/2$$

character comparisons are needed. The run time of naïve suffix tree construction is therefore proportional to $n(n - 1)/2$, which scales as n^2 or $O(n^2)$. This is similar to the run time of optimal alignment, that is, too slow for genomics.

Answer 201 Use commands like

```
ranseq -l 1000000 | time repeater
```

Collect the results in `time.dat` and plot them using

```
gnuplot -p stTime.gp
```

where `stTime.gp` is

```
set xlabel "Sequence Length (Mb)"
set ylabel "Run Time (s)"
plot "time.dat" title "" with lines
```

to get

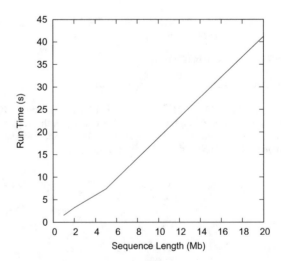

For sequences of length n, the run time of `repeater` is $O(n)$. This linear run time behavior is optimal in the sense that it cannot be improved upon. The algorithm implemented in `repeater` was devised in 1995 [48] and is described in detail in a classic textbook of bioinformatics [21].

Answer 202 Say, `memory.dat` contains the memory measurements. Then

```
gnuplot -p stMemory.gp
```

where `stMemory.gp` is

```
set xlabel "Sequence Length (Mb)"
set ylabel "Memory (MB)"
plot "memory.dat" title "" with lines
```

returns

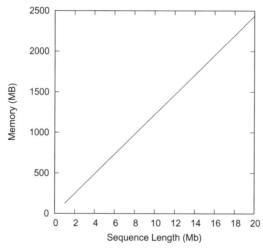

Like time, memory consumption is linear in sequence length. With almost 2.5 GB for 20 Mb of sequence, it is quite large, though.

Answer 203 The command

```
awk -f suf.awk s.fasta | sort -k 2 | cat -n
```

gives the same suffix array as in Fig. 3.6a, minus the header.

Answer 204 The root corresponds to sa[1..6], the node with path label C to sa[4..6], and the node with path label CC to sa[5..6].

Answer 205 Here are the common prefixes:

index	sa	suf	cp
1	6	$	nd
2	5	A$	–
3	1	ACCCA$	A
4	4	CA$	–
5	3	CCA$	C
6	2	CCCA$	CC

The color codes for the common prefixes are repeated in the suffix tree:

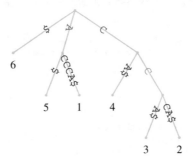

All three edges leading to an inner node of the suffix tree are labeled by a common prefix, because a suffix tree essentially summarizes the common prefixes of all suffixes.

Answer 206

index	sa	suf	cp	lcp
1	6	$	nd	-1
2	5	A$	–	0
3	1	ACCCA$	A	1
4	4	CA$	–	0
5	3	CCA$	C	1
6	2	CCCA$	CC	2

As the first entry in cp is undefined (there is no suffix "above" it), we give it a length less than the smallest entry in the lcp array; by convention −1 is used.

Answer 207 The remaining lcp intervals are as follows:

	(e)		2 1 0		(f)		2 1 0		(g)		2 1 0
index	sa	lcp		index	sa	lcp		index	sa	lcp	
1	6	-1		1	6	-1		1	6	-1	
2	5	0		2	5	0		2	5	0	
3	1	1		3	1	1		3	1	1	
4	4	0		4	4	0		4	4	0	
5	3	1		5	3	1		5	3	1	
6	2	2		6	2	2		6	2	2	
7	–	-1		7	–	-1		7	–	-1	

Answer 208

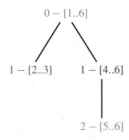

$$0 - [1..6]$$
$$1 - [2..3] \quad 1 - [4..6]$$
$$2 - [5..6]$$

This lcp interval tree is the suffix tree in Fig. 3.6 stripped of its leaves.

Answer 209 On the left is the lcp interval tree, on the right the suffix tree.

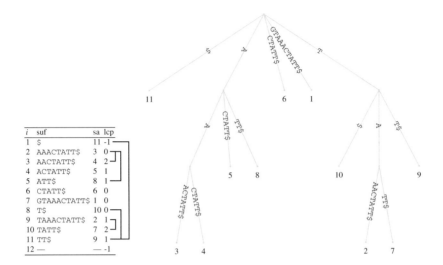

i	suf	sa	lcp
1	$	11	-1
2	AAACTATT$	3	0
3	AACTATT$	4	2
4	ACTATT$	5	1
5	ATT$	8	1
6	CTATT$	6	0
7	GTAAACTATT$	1	0
8	T$	10	0
9	TAAACTATT$	2	1
10	TATT$	7	2
11	TT$	9	1
12	—	—	-1

Answer 210

isa i	sa	suf	cp	lcp
3 1	6	$	–	-1
6 2	5	A$	–	0
5 3 → 1	ACCCA$	A	1
4 4	4	CA$	–	0
2 5	3	CCA$	C	1
1 6 → 2	CCCA$	CC	2

By following the arrows, convince yourself that sa[isa[1]] = 1, sa[isa[2]] = 2, and so on. Thus by traversing the inverse suffix array isa, the suffixes in the alphabetically sorted suffix array sa are visited in original order.

Answer 211

```
{
    n = n + 1        # position in the suffix array
    sa[n] = $1       # save the suffix array entries
    suf[n] = $2      # save the suffixes
    isa[sa[n]] = n # construct inverse suffix array
} END {
    printf "# i\tsa\tisa\tsuf\n" # print the output table
    for(i=1; i<=n; i++)
        printf "%d\t%d\t%d\t%s\n", i, sa[i], isa[i], suf[i]
}
```

Run it

```
awk -f suf.awk s.fasta | sort -k 2 | awk -f isa.awk
# i     sa        isa      suf
1       6         3        $
2       5         6        A$
3       1         5        ACCCA$
4       4         4        CA$
5       3         2        CCA$
6       2         1        CCCA$
```

Answer 212

```
{
    n++                 # position in the suffix array
    sa[n] = $1          # save the suffix array
    suf[n] = $2         # save the suffixes
    if(sa[n] == 1)      # find the input sequence
        s = $2
}END{
    # compute the isa
    for(i=1; i<=n; i++)
        isa[sa[i]] = i
    # compute the lcp-array
    lcp[1] = -1
    L = 0
    for(i=1; i<=n; i++){
        j = isa[i]
        if(j > 1){
            k = sa[j-1]
            while(substr(s, i+L, 1) == substr(s, k+L, 1))
                L++
            lcp[j] = L
            if(L > 1)
                L = L - 1
            else
                L = 0
        }
    }
    # print the output table
    printf "# i\tsa\tlcp\tsuf\n"
    for(i=1; i<=n; i++)
        printf "%d\t%d\t%d\t%s\n", i, sa[i], lcp[i], suf[i]
}
```

This can be run as

```
awk -f suf.awk s.fasta | sort -k 2 | awk -f esa.awk
# i     sa        lcp      suf
1       6         -1       $
2       5         0        A$
3       1         1        ACCCA$
4       4         0        CA$
5       3         1        CCA$
6       2         2        CCCA$
```

Answer 213

```
awk -f suf.awk dgAdhAdhdup.fasta  |
sort -k 2                         |
awk -f esa.awk                    |
sed '/^#/d'                       |
sort -k 3 -n -r                   |
head -n 1
3325    988 12         TACATTACATTA...
```

The longest repeat has length 12, and it is found at positions 988 and 3325.

Answer 214

```
awk -f suf.awk dmAdhAdhdup.fasta  |
sort -k 2                         |
awk -f esa.awk                    |
sed '/^#/d'                       |
sort -k 3 -n -r                   |
head -n 1
3890    3908    16        TCGATGTCCTGATCAA...
```

The longest repeat has length 16, and it is found at positions 3890 and 3908.

Answer 215

```
awk -f suf.awk dmDgAdhAdhdup.fasta  |
sort -k 2                           |
awk -f esa.awk                      |
sed '/^#/d'                         |
sort -k 3 -n -r                     |
head -n 1
1592    8048    37
        AGCAAGGTTCTCATGACCAAGAATATAGCGGTGAGTG...
```

The longest repeat for the concatenated sequences must be at least as long as the longer of the two repeats seen in the individual sequences, that is, it has to be at least 16 bp long. However, the two dehydrogenase sequences were taken from two *Drosophila* species; the relatedness between these species means there is much more sequence similarity between than within the alcohol sequences, leading to the much longer repeat of 37 bp.

Answer 216

Position	Shustring
1	AC
2	CCC
3	CCA
4	CA
5	nonexistent

The shortest shustrings are AC and CA with length 2. In the context of real sequences, shustrings could be used for the design of PCR primers, or in DNA-based identification of pathogens.

Answer 217 The enhanced suffix array for ACCCA is shown in the Solution to Problem 206—or you could generate it using suf.awk and esa.awk. To find the shustring for a given position, i, look at lcp[i] and the following entry, lcp[$i + 1$]. Whichever is greater is the length of the longest repeat starting at sa[i]. This value plus one gives the shustring length at sa[i].

Answer 218

```
shustring -i mgGenome.fasta
#>gi|84626123|gb|L43967.2| Mycoplasma genitalium G37,
    complete genome 580076 35 6
#num  pos       len  seq
1     11911     6    CGAGGC
2     37969     6    GAGACG
3     60188     6    TCGGAC
...
```

So there are 35 unique motifs of length 6.

Answer 219 Including the reverse strand gives

```
shustring -r -i mgGenome.fasta
#>gi|84626123|gb|L43967.2| Mycoplasma genitalium G37,
    complete genome 580076 2 6
#num  pos       len  seq
1     174222    6    GACGGC
2     567107    6    GCCGGG
```

In other words, of the 35 shustrings found on the forward strand, 33 also occurred on the reverse strand, leaving only two shustrings of length 6 when scanning the whole genome.

Answer 220

```
shustring -l . -M 7 -r -i mgGenome.fasta |
head -n 1
#>L43967 L43967.1 Mycoplasma genitalium G37 complete genome.
    580074 254 6<=l<=7
```

So there 254 shustrings $\leq$ 7 bp long.

Answer 221

Rotations	Sort
TACTA$	$TACTA
ACTA$T	A$TACT
CTA$TA	ACTA$T
TA$TAC	CTA$TA
A$TACT	TA$TAC
$TACTA	TACTA$

The BWT is the last column of the sorted rotations:

```
ATTAC$
```

Make the directory for this session and change into it

```
mkdir Bwt
cd Bwt
```

Check the transform

```
bwt seq.fasta
```

where `seq.fasta` contains

```
>Seq
TACTA
```

Answer 222 We find long runs of identical letters, for example

```
...
115 LLLLLLLLLLLLLLLLLLLLLLLLLLLLLLLLLLLLLLLLLLLLLLLLLLLLLLLLLLLLLLLLNNNNNNN
116 NNNNNNNNNNNNNRHHHHHHHHHHHHHHHHHHHHHHHHHHHHHHHHHHHHHHHHHHHHHHHHHHHHHHHHHH
117 HHHHHHHHHHHHHHHHHHHHHHHHHHHHHHHHHHHHHHHHHHHHHHHHHHHHHHHHHHHHHHHHHHHHHHHH
118 HHHHHHHHHHHHHHHHHHHHHHHHHHHHHHHHHHHHHHHHHHHHHHHHHHHHHHHHHHHHHHHHHHHHHHHH
119 HHHHHHHHHHHHHHHHHHHHHHHHHHHHHHHHHHHHHHHHHHHHHHHHHHHHHHHHHHHHHHHHHHHHHHHH
120 HHHHHHHHHHHHHHHHHHHHHHHHHHHHHHHHHHHHHHHHHHHHHHHHHHHHHHHHHHHHHHHHHHHHHHHH
121 HHHHHHHHHHHHHHHHHHHHHHHHHHHHHHHHHHHHHHHHHHHHHHHHHHHHHHHHHHHHHHHHHHHHHHHH
122 HRRRRRRRRRRMMMMMMMRRRRRRRRRRRRRRRRRRRRRRRRRRRRRRRRRRRRRRRRRRRRRRRRRRRRRR
...
```

Answer 223 The BWT is the last column of the sorted rotations. For decoding, we also need the first column of the rotation, which we obtain by sorting the transform:

Sort		Count		Reconstruct	
$\mathcal{F}$	$\mathcal{L}$	$\mathcal{F}$	$\mathcal{L}$	$\mathcal{F}$	$\mathcal{L}$
$	C	$_1$	C$_1$	$_1$	C$_1$
C	C	C$_1$	C$_2$	C$_1$	C$_2$
C	C	C$_2$	C$_3$	C$_2$	C$_3$
C	T	C$_3$	T$_1$	C$_3$	T$_1$
G	$	G$_1$	$_1$	G$_1$	$_1$
T	G	T$_1$	G$_1$	T$_1$	G$_1$

For the decoding, we trace back from $_1$ in $\mathcal{L}$ to $_1$ in $\mathcal{F}$ and note the nucleotides in $\mathcal{F}$ along that path, which returns GTCCC. To check this result, enter

```
bwt -d bwt.fasta
```

where `bwt.fasta` contains CCCT$G.

Answer 224 The key insight is that the last character in the rotation is just to the left of the start of a suffix. Hence, the BWT of a string is found by looking up $T[\text{sa}[i] - 1]$ for $i = 1, 2, \ldots$. The only exception comes when sa[i] is 1: The character to the left of $T[1]$ by definition is the sentinel.

Answer 225 We write down our text with indexes:

1 2 3 4 5 6
T A C T A $

Then we construct its suffix array:

i	sa	suf
1	6	$
2	5	A$
3	2	ACTA$
4	3	CTA$
5	4	TA$
6	1	TACTA$

Finally, write down $T[sa[i]-1]$ for $i = 1, ..., 6$ to get bwt$(T) =$ ATTAC$. Check the result

```
bwt seq.fasta
ATTAC$
```

where seq.fasta contains TACTA.

Answer 226

	0 1 2 3
Encoding	A C G T
2	G A C T
2,3	T G A C
2,3,0	T G A C
2,3,0,2	A T G C
2,3,0,2,2	G A T C

So the solution is 2,3,0,2,2. To check, enter

```
mtf mtf.fasta
```

where mtf.fasta contains GTTAG.

Answer 227

	0 1 2 3
Decoding	A C G T
C	C A G T
CC	C A G T
CCA	A C G T
CCAT	T A C G
CCATG	G T A C

So the solution is CCATG.

Answer 228 The decomposition is

```
T.A.C.TA
```

which we can verify using

```
lzd seq.fasta
T.A.C.TA
```

Answer 229 This microsatellite decomposes into two factors:

```
A.AAAA
```

In fact, all sequences of a single kind of nucleotide decompose into just two factors, regardless of their length.

Answer 230 The sequence TACTA contains 4 LZ factors, and hence its complexity is $4/5 = 0.8$, while AAAA decomposes into two factors, so its complexity is $2/5 = 0.4$.

Answer 231 The smallest number of factors is 2, and hence $C \geq 2/|S|$. The largest number of factors is equal to the sequence length, so

$$2/|S| \leq C \leq 1.$$

Answer 232 We carry out the computations

```
for a in 1 2 5 10; do ranseq -l ${a}000 | lzd -n; done
# n    n/site
247    0.246753
# n    n/site
441    0.22039
# n    n/site
964    0.192761
# n    n/site
1751   0.175082
```

to get

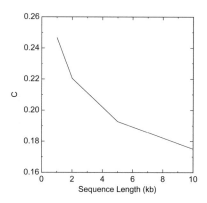

We see that the maximum complexity, C, depends on the sequence length. So we can only sensibly compare C between sequences of the same length.

Answer 233 We use commands like

```
bwt mgGenome.fasta | lzd -n
```

to get

Operation	Number of Factors	C
None	60049	0.104
BWT	62688	0.108
BWT | MTF	45408	0.060
MTF	63495	0.109
MTF | BWT	65261	0.113

There is only one combination that results in any complexity reduction, BWT followed by MTF.

Answer 234 Measure the size of the original file

```
du -h mgGenome.fasta
576K
```

Then, measure the sizes of the transformed and compressed files to get

Operation	gzip	bzip2
nothing	172	160
randomizeSeq	176	164
BWT	176	164
BWT | MTF	124	116

As before, the combination of BWT and MTF leads to the greatest compressibility.

Answer 235 We begin as usual by creating a working directory:

```
mkdir KerrorAlignment
cd KerrorAlignment
```

Then copy the sequence

```
cp ../Data/dmAdhAdhdup.fasta .
```

and cut out the desired region

```
cutSeq -r 2301-2400 dmAdhAdhdup.fasta > dmAdhFrag.fasta
```

Check the result by confirming it matches its original position:

```
keywordMatcher -f dmAdhFrag.fasta dmAdhAdhdup.fasta
>DMADH X78384.1 ... :2301
```

Answer 236 Mutate the sequence:

```
mutator -p 10 dmAdhFrag.fasta > dmAdhFrag2.fasta
```

Check the mutation:

```
gal -i dmAdhFrag.fasta -j dmAdhFrag2.fasta
```

Confirm the exact match is gone:

```
keywordMatcher -f dmAdhFrag2.fasta dmAdhAdhdup.fasta
```

However,

```
lal -i dmAdhFrag2.fasta -j dmAdhAdhdup.fasta
```

works as expected.

Answer 237 Locate the error-free copy

```
kerror -i dmAdhFrag.fasta -j dmAdhAdhdup.fasta
```

and the mutated copy with the fragments printed

```
kerror -L -i dmAdhFrag2.fasta -j dmAdhAdhdup.fasta
```

Now search for the fragments; the first one is not found,

```
keywordMatcher -p
   GACCACCAACCTGCTGAAGACCATCTTCGCCCAGCTGAAGACCGTCGATG
   dmAdhAdhdup.fasta
```

but the second one is

```
keywordMatcher -p
   TCCTGATCAACGGAGCTGGTATCCTGGACGATCACCAGATCGAGCGCACC
   dmAdhAdhdup.fasta
```

Answer 238 The alignment produced by `kerror` is global in the query and local in the subject. Such "glocal" alignments are used to align sequencing reads to genomes.

Answer 239 Copy the chromosome files

```
cp ../Data/dmChr*.fasta .
```

Then list their sizes:

```
cchar -s dmChr*.fasta | grep '^>'
```

to get

```
>NT_033779.5 Drosophila melanogaster chromosome 2L 23513712
>NT_033778.4 Drosophila melanogaster chromosome 2R 25286936
>NT_037436.4 Drosophila melanogaster chromosome 3L 28110227
>NT_033777.3 Drosophila melanogaster chromosome 3R 32079331
>NC_004353.4 Drosophila melanogaster chromosome 4 1348131
>NC_024511.2 Drosophila melanogaster mitochondrion, complete
    genome 19524
>NC_004354.4 Drosophila melanogaster chromosome X 23542271
>NC_024512.1 Drosophila melanogaster chromosome Y 3667352
```

Compute the complete genome length:

```
cchar -s dmChr*.fasta      |
grep '^>'                  |
cut -f 2                   |
awk '{s+=$1}END{print s}'
137567484
```

The genome is approximately 138 Mb long.

Answer 240 Copy *Hamlet*

```
cp ../Data/hamlet.fasta .
```

and take a look at it

```
less hamlet.fasta
```

before counting its characters

```
cchar hamlet.fasta | head -n 1
# Total number of input characters: 136033
```

This means that the genome of *D. melanogaster* is approximately 1000 times longer than the text of *Hamlet*.

Answer 241 Compare the links and the original files using commands like

```
diff dmChr2L.fasta ../Data/dmChr2L.fasta
```

to find that content-wise they are identical. However, their sizes are vastly different: The command `ls -l` returns entries like

```
ls -l
lrwxrwxrwx 1 haubold haubold   27 Apr 17 16:01 dmChr2L.fasta
    -> ../Data/dmChr2L.fasta
```

which indicates the target of the symbolic link, `->`, and its size, 27 bytes. The original file, on the other hand, is almost a million times larger:

```
ls -l ../Data/dmChr2L.fasta
-rw-rw-r-- 1 haubold haubold 23807685 Apr 15 09:59
    ../Data/dmChr2L.fasta
```

Answer 242 Run a script like

```
for k in 1 2 5 10 20 50 100 200 500
do
    echo $k
    for a in 2L 2R 3L 3R 4 X Y Mt
    do
        kerror -k $k -i dmAdhAdhdup.fasta -j dmChr${a}.fasta
    done
done
```

It prints an alignment where *Adh/Adh-dup* is located on the left arm of chromosome 2 at positions 14,614,315–14,619,086; the alignment contains 161 errors, which is a surprisingly large number given that both sequences were sampled from the same species.

Answer 243 The number of errors per site is

$$\pi = \frac{161}{14,619,086 - 14,614,315 + 1} \approx 3.4\%.$$

Answer 244 Most errors are located toward the end of the alignment.

Answer 245 Cut out the *Adh/Adh-dup* locus:

```
cutSeq -r 14614315-14619086 dmChr2L.fasta > dmGenomic.fasta
```

Align the sequences

```
lal -i dmGenomic.fasta -j dmAdhAdhdup.fasta > aln.lal
```

This matches positions 1–4597 in the query and 1–4589 in the subject. Count the matches

```
grep '|' aln.lal    |                 # Extract match lines
sed 's/ *//g'       |                 # Remove blanks
awk '{s+=length($1)}END{print s}'   # Count match symbols
4541
```

This means that the mismatches per site is

$$\pi = \frac{4589 - 4541}{4589} \approx 1\%.$$

Answer 246 The command

```
kerror -k 161   -i dmAdhAdhdup.fasta -j dmChr2L.fasta
```

gives as time measurements on our computer

```
# Total time:          3.42s
# Data manipulation:  0.14s
# Matching:            0.49s
# Checking:            2.80s
```

This means that matching the $k + 1 = 162$ fragments is five times faster than checking whether or not they are part of a full alignment.

Answer 247 The *Adh/Adh-dup* region is already contained in dmGenomic.fasta. Align it to the *D. guanche* version:

```
gal -i dgAdhAdhdup.fasta -j dmGenomic.fasta > aln.gal
```

Count the matches

```
grep '|' aln.gal    |
sed 's/ *//g'       |
awk '{s+=length($1)}END{print s}'
2960
```

Since dmGenomic.fasta is 4772 bp long, kerror needs to be run with $k = 4772 - 2960 = 1812$. In other words, dgAdhAdhdup.fasta, which is 4433 bp long, is divided into fragments of length $4433/(1812+1) \approx 2.45$. Since matches of this length are ubiquitous, the checking phase would take a very long time and make the search unfeasible in practice.

Answer 248 Create the directory for this session, change into it, and copy the input sequence:

```
mkdir FastLocalAlignment
cd FastLocalAlignment
cp ../Data/dmAdhAdhdup.fasta .
```

Cut out the fragment:

```
cutSeq -r 3101-3200 dmAdhAdhdup.fasta > dmAdhFrag.fasta
```

Align it to the original sequence:

```
sblast -i dmAdhFrag.fasta -j dmAdhAdhdup.fasta
# reading input data...done
# step1: generating word list from query...done
# step2: searching for exact matches of words in subject...done
# step3: extending exact matches...done
# qs    qe   ss    se    score
1       100  3101 3200  100.0
```

The result is exactly at the expected position. Repeat the alignment with the word list printed out:

```
sblast -L -i dmAdhFrag.fasta -j dmAdhAdhdup.fasta
```

There are 90 words, each 11 bp long.

Answer 249 Copy the fragment

```
cp dmAdhFrag.fasta dmAdhFrag2.fasta
```

Mutate it

```
bash mutate.sh
```

where mutate.sh is

```
for i in $(seq 1 11 100)
do
    mutator -p ${i} dmAdhFrag2.fasta > tmp
    mv tmp dmAdhFrag2.fasta
done
```

Then align it using sblast:

```
sblast -i dmAdhFrag2.fasta -j dmAdhAdhdup.fasta
```

where no hit is found. This is because there is no word without a mismatch and hence no starting point for the extension step (Fig. 4.2c).

Answer 250 Any word length less than 11 will return an alignment, for example,

```
sblast -w 10 -i dmAdhFrag2.fasta -j dmAdhAdhdup.fasta
```

Answer 251 The command

```
lal -i dmAdhFrag2.fasta -j dmAdhAdhdup.fasta
```

returns the expected alignment. In other words, `lal` is more sensitive than `sblast` with default parameters.

Answer 252 Run

```
bash sensitivitySblast.sh > sensitivity1.dat
gnuplot -p sensitivity1.gp
```

where `sensitivitySblast.sh` is

```
for m in 0.01 0.02 0.05 0.1 0.2 0.5
do
    echo -n ${m} ' '
    for a in $(seq 100)
    do
        mutator -m ${m} dmAdhFrag.fasta > dmAdhFrag3.fasta
        sblast -i dmAdhFrag3.fasta -j dmAdhAdhdup.fasta
    done |
        grep -A 1 qs | sed '/^#/d;/^-/d' | wc -l
done
```

Instead of the line

```
grep -A 1 qs | sed '/^#/d;/^-/d' | wc -l
```

you might have used

```
grep -v ^# | wc -l
```

which is simpler and gives almost the same result. However, the more complex filter ensures that no more than one result is counted per `sblast` run. Compare the two solutions without `wc -l` to see the difference. The gnuplot script `sensitivity1.gp` contains

```
set logscale x
set xlabel "Mutation Rate"
set ylabel "% Alignments Found"
plot "sensitivity1.dat" title "" with lines
```

The resulting plot is

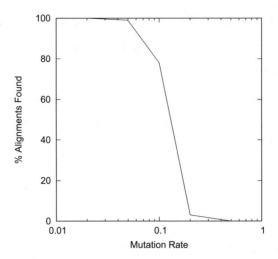

There is a sharp drop in sensitivity for queries mutated at more than 20% of their positions.

Answer 253 In `sensitivitySblast.sh` change the line

```
sblast -i dmAdhFrag3.fasta -j dmAdhAdhdup.fasta
```

to

```
sblast -t 25 -i dmAdhFrag3.fasta -j dmAdhAdhdup.fasta
```

Save the simulation results in `sensitivity2.dat` and plot

```
gnuplot -p sensitivity2.gp
```

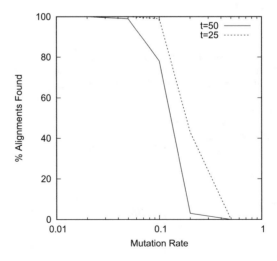

where `sensitivity2.gp` is

```
set xlabel "Mutation Rate"
set ylabel "% Alignments Found"
set logscale x
plot "sensitivity1.dat" title "t=50" with lines,\
"sensitivity2.dat" title "t=25" with lines
```

to see that the sensitivity of `sblast` is increased if the minimum score is halved from 50 to 25.

Answer 254

```
sblast -i dmAdhAdhdup.fasta -j dgAdhAdhdup.fasta |
sed '/^#/d'                                       |
sort -k 5 -n -r
2299     2594     2259     2554     164.0
3865     4028     3657     3820     76.0
3225     3328     3220     3323     68.0
2182     2285     2142     2245     64.0
```

The optimal local alignment is found by

```
lal -i dmAdhAdhdup.fasta -j dgAdhAdhdup.fasta
```

It has a score of 217, which is much larger than the score of the best alignment found by `sblast` (164).

Answer 255 The command we are looking for is

```
sblast -s 40 -i dmAdhAdhdup.fasta -j dgAdhAdhdup.fasta
# reading input data...done
# step1: generating word list from query...done
# step2: searching for exact matches of words in subject...done
# step3: extending exact matches...done
# qs      qe       ss       se       score
3829     4028     3621     3820     80.0
3225     3328     3220     3323     68.0
2182     2594     2142     2554     217.0
```

Now the best alignment has the same coordinates and the same score as the best alignment returned by `lal`. Heuristics like the extension parameter can influence the result.

Answer 256 Run

```
bash driveSblastDm.sh dmChr*.fasta
```

where `driveSblastDm.sh` contains

```
for a in $@
do
    echo Searching ${a}
    sblast -i dmAdhAdhdup.fasta -j $a |
        sed '/^#/d'
done
```

to see—as we have already done with `kerror`—that in *D. melanogaster* the *Adh/Adh-dup* region is located on the left arm of chromosome 2. To find the exact interval, run

```
bash driveSblastDm.sh dmChr*.fasta |
grep ^[0-9]                          |
sort -n -k 3
```

to locate *Adh/Adh-dup* in the interval 14,614,315–14,619,393.

Answer 257

```
time kerror -k 161 -i dmAdhAdhdup.fasta -j dmChr2L.fasta
```

takes 3.79 s, while

```
time sblast -i dmAdhAdhdup.fasta -j dmChr2L.fasta
```

takes only 1.37 s. The program `sblast` is faster than `kerror`, because `kerror` uses dynamic programming for finding the alignment (Fig. 4.1), while `sblast` uses the simpler extension strategy (Fig. 4.2). Apart from speed, `sblast` has the advantage that there is no need for guessing a suitable number of errors, k.

Answer 258 Run

```
sblast -i dgAdhAdhdup.fasta -j dmChr2L.fasta |
grep ^[0-9]                                   |
sort -n -k 3
```

to find homology in the interval 14,616,500–14,618,356. This shows that even ungapped alignment can be a highly effective tool: it is faster and more sensitive than `kerror`.

Answer 259 The command

```
blastn -query dmAdhFrag2.fasta -subject dmAdhAdhdup.fasta
```

returns no hit. However, with the appropriate -word_size the expected hit is found, plus three more of low significance:

```
blastn -query dmAdhFrag2.fasta -subject dmAdhAdhdup.fasta -
    word_size 10
```

Answer 260 Our manual binary search resulted in the following left and right borders (l & r), where the middle, $m = (l + r)/2$:

Step	l	r	m
1	1	100	51
2	1	51	26
3	26	51	39
4	26	39	33
5	26	33	30
5	26	30	28
6	28	30	29

Since we got a hit with a distance of 29 between mutations, but not with a distance of 28, the default value of -word_size is 28.

Answer 261 Here are the steps of the binary search:

Step	l	r	m
1	1	100	51
2	1	51	26
3	1	26	14
4	1	14	8
5	8	14	11
5	11	14	13
6	11	13	12

Since we got a hit with a distance of 12 between mutations, but not with a distance of 11, the default -word_size is now 11.

Answer 262 To explore the sensitivity in blastn mode, we simulate

```
bash sensitivityBlastn.sh > sensitivityN.dat
```

where sensitivityBlastn.sh is

```
for m in 0.01 0.02 0.05 0.1 0.2 0.5
 do
    echo -n ${m} ' '
    for i in $(seq 100)
    do
        mutator -m ${m} dmAdhFrag.fasta > dmAdhFrag3.fasta
        blastn -outfmt 7 -task blastn -query dmAdhFrag3.
            fasta -subject dmAdhAdhdup.fasta
    done  |
        grep -A 1 hits | sed '/^#/d;/^-/d' | wc -l
done
```

For the megablast mode, just leave out

```
-task blastn
```

from the blastn command and run

```
bash sensitivityMega.sh > sensitivityM.dat
```

Plot the two result sets

```
gnuplot -p sensitivity3.gp
```

where sensitivity3.gp is

```
set xlabel "Mutation Rate"
set ylabel "% Alignments Found"
set logscale x
set key bottom left
plot "sensitivityM.dat" title "megablast" with lines,\
"sensitivityN.dat" title "blastn" with lines
```

We moved the key from the top right to the bottom left—otherwise it would intersect the graph in our printout; the blastn mode is much more sensitive than the megablast mode.

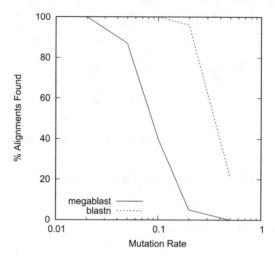

Answer 263 All that needs to be changed is the match score in lal:

```
lal -A 2 -i dmAdhFrag2.fasta -j dmAdhAdhdup.fasta
```

This returns the same alignment as blastn with the same score of 156.

Answer 264 The command

```
blastn -reward 1 -task blastn -query dmAdhFrag2.fasta
   -subject dmAdhAdhdup.fasta
```

returns the match line

```
Score = 131 bits (66),   Expect = 1e-34
```

in which the bit score has changed only from previously 141 to 131, but the raw score has been more than halved from 156 to 66.

Answer 265 The best alignment with the BLAST command

```
blastn -task blastn -subject dmAdhAdhdup.fasta -query
   dgAdhAdhdup.fasta
```

has a raw score of 842. In contrast, the best alignment with lal,

```
lal -A 2 -i dmAdhAdhdup.fasta -j dgAdhAdhdup.fasta
```

has the much higher score of 1088.

Answer 266 A bit of trial and error gave 333 as the smallest xdrop value compatible with the alignment returned by lal:

```
blastn -task blastn -query dmAdhAdhdup.fasta -subject
   dgAdhAdhdup.fasta -xdrop_gap_final 333
```

Answer 267 `Blastn` cannot be used to find a better alignment than `lal`. Remember, optimal alignment methods are guaranteed to return the best result given a score scheme.

Answer 268 Cut out the desired region

```
cutSeq -r 3101-3200 dgAdhAdhdup.fasta > dgAdhFrag.fasta

blastn -task blastn -query dgAdhFrag.fasta -subject
   dmAdhAdhdup.fasta
```

gives an alignment with raw score 22. However, the corresponding `lal` command

```
lal -A 2 -i dgAdhFrag.fasta -j dmAdhAdhdup.fasta
```

gives a better alignment with score 26. With a bit of trial and error, we found 7 is the largest word size compatible with the optimal alignment:

```
blastn -word_size 7 -task blastn -query dgAdhFrag.fasta
   -subject dmAdhAdhdup.fasta
```

Answer 269

$$P = 1 - e^{-0.015} \approx 0.015,$$

which illustrates that $P \leq E$, and also that for small values of E the two statistics are quite similar. Beware, however, P is a probability and hence bounded by 0 and 1, while E is an expectation value with lower bound 0 but no obvious upper bound.

Answer 270 The desired parameters are as follows:

Name	Value
s	413076
λ	0.625
K	0.410

We can use AWK to calculate

```
BEGIN{
    s = 413076
    l = 0.625
    K = 0.41
    x = 26
    y = K*s*exp(-l*x)
    print 1-exp(-y)
}
```

and get $P \approx 0.015$, which is exactly the value implied by the E-value returned.

Answer 271 The script

```
time bash simPval.sh > simPval.dat
```

where simPval.sh contains

```
for a in $(seq 1000)
do
      randomizeSeq dmAdhAdhdup.fasta   |    # randomize subject
          lal -A 2 -i dgAdhFrag.fasta |    # alignment
          grep '^Sc'                   |    # extract score line
          sed 's/Score: *//'                # extract score
done
```

takes 98 s to execute. So if we decided to run the simulation for 10^4 iterations, we would have to wait approximately 17 min.

Answer 272 Count the results

```
awk -f count.awk simPval.dat
```

where count.awk is

```
{
    arr[$1]++
    c++
}END{
    for(a in arr)
        print a "\t" arr[a]/c
}
```

and plot them

```
awk -f count.awk simPval.dat |
sort -n                      |
gnuplot -p simPval1.gp
```

where simPval1.gp is

```
set xlabel "Score"
set ylabel "Frequency(Score)"
set arrow from 26,0.05 to 26,0.03
plot "< cat" title "" with lines
```

to get the distribution of random scores

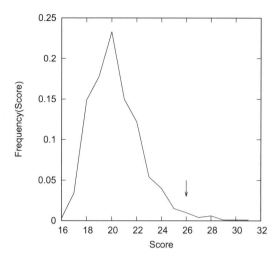

where the arrow points to the observed score.

Answer 273 The P value is

```
awk '{if($1>=26)s++;c++}END{print s/c}' simPval.dat
0.022
```

Your result is bound to differ slightly. However, the simulated P value should always be at least similar to the theoretical $P = 0.015$.

Answer 274 Plot

```
awk -f count.awk simPval.dat  |
sort -n                       |
gnuplot -p simPval2.gp
```

where `simPval2.gp` is

```
s=413076
l=0.625
K=0.410
mu=log(K*s)/l
f(x)=l*exp((mu-x)*l-exp((mu-x)*l))
set xlabel "Score"
set ylabel "P(Score)"
set arrow from 26,0.05 to 26,0.03
plot "< cat" title "simulated" with lines,\
f(x) title "theoretical" with lines
```

to get

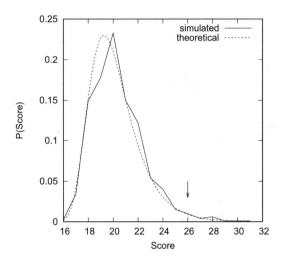

Again, the arrow points to the observed score. Notice that we now interpret the frequency of a score as its probability, and hence the label $P(\text{Score})$ along the y-axis. The significance of the observed score of 26 (its P-value) is the proportion of random scores ≥ 26 obtained either from the theoretical or the simulated curve. Since the two curves agree well, and the simulations are time-consuming, calculating the E-value directly is by far the better option.

Answer 275 The search

```
time bash blast.sh dmChr*.fasta
```

where `blast.sh` is

```
for a in $@
do
    echo $a
    blastn -task blastn -query dmAdhAdhdup.fasta -subject $a
        -evalue 1e-20 -outfmt 7
done
```

takes approximately 5 s and returns two alignments covering positions 14,614,315–14,618,911 and 14,619,298–14,619,405.

Answer 276 Database construction takes approximately 2 s. The search based on this database

```
time blastn -task blastn -outfmt 7 -query dmAdhAdhdup.fasta
    -db dmDb -evalue 1e-20
```

takes 0.7 s. So the combined run time of database construction plus search is 2.7 s, which is less than the 5 s it took to search the subject files in plain FASTA format.

Answer 277 The run

```
time blastn -outfmt 7 -query dmAdhAdhdup.fasta -db dmDb
    -evalue 1e-20
```

takes 0.15 s, so it is approximately five times faster than the blastn mode. The two alignments found have the same start and end coordinates as those found in blastn mode. This is because we are carrying out a sequence comparison within a species, that is, between very similar sequences, for which megablast is optimized. However, the alignments themselves do differ slightly; for example, in the first alignment we found:

Parameter	blastn	megablast
alignment length	4601	4603
mismatches	44	39
gap opens	10	15
bit score	7997	8155

Answer 278 Now the results do differ. In blastn mode, we get three alignments with a combined length of $169 + 173 + 37 = 379$ bp. The 169 bp alignment is the left-most hit, starting at position 14,616,354; the 173 bp alignment is the right-most hit, ending at position 14,618,839. Taken together, these three alignments cover the 2485 bp interval 14,616,354–14,618,839. In contrast, the megablast mode returns a single alignment of length 134 located at 14,617,517–14,617,650. Notice also the difference in E-values: In blastn mode, these are 0, 3×10^{-159}, and 10^{-37}; in megablast mode, the single alignment has $E = 3 \times 10^{-33}$.

Answer 279 Here is the overlap between the two example sequences:

```
ACCGTTC----
---GTTCAGTA
```

Answer 280 Create the directory, change into it, and generate the sequence files:

```
mkdir Shotgun
cd Shotgun
echo -e '>S1\nACCGTTC'  > s1.fasta
echo -e '>S2\nGTTCAGTA' > s2.fasta
```

Now compute the overlap alignment

```
oal -i s1.fasta -j s2.fasta
```

```
Query:              >s1
   Length:          7
Subject:            >s2
   Length:          8
Score:              4.0
Strand:             Plus / Plus


Query: 1 ACCGTTC---- 7
         ||||
Sbjct: 1 ---GTTCAGTA 8
//
```

Answer 281 You might be tempted to think that four comparisons are necessary: forward/forward, forward/reverse, reverse/forward, and reverse/reverse. But forward/forward is equivalent to reverse/reverse and forward/reverse is equivalent to reverse/forward, so only two comparisons are necessary.

Answer 282 Create the forward files

```
for a in $(seq 3); do cp f${a}.fasta f${a}f.fasta; done
```

Create the reverse files

```
for a in $(seq 3); do revComp f${a}.fasta > f${a}r.fasta; done
```

Then carry out the comparisons like

```
oal -i f1f.fasta -j f2f.fasta | grep Score
```

to find the following scores

	f_1^f	f_1^r	f_2^f	f_2^r	f_3^f	f_3^r
f_1^f		2	16	0	30	
f_1^r						
f_2^f				1	0	
f_2^r						
f_3^f						
f_3^r						

The two most substantial overlaps are between f1 and f2, and between f1 and f3. This means f1 bridges f2 and f3.

Answer 283 There are several ways of solving this, here is ours: Begin with `f1f.fasta` and `f2r.fasta` and get an overlap alignment like this

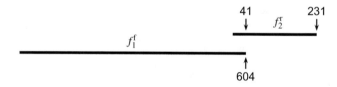

where 41 is the last overlapping position in f_2^r. Then, compare `f1f.fasta` and `f3r.fasta`, and add the next overlap, where 197 is the first overlapping position in f_3^r:

Now we can calculate the length of the underlying genomic region as

$$196 + 604 + 190 = 1000.$$

The reads were drawn from a 1 kb fragment.

Answer 284 Compute the genome size of *M. genitalium*:

```
cchar mgGenome.fasta
```

```
# Total number of input characters: 580076
# Char   Count    Fraction
A        200544   0.345720
C        91515    0.157764
G        92306    0.159127
T        195711   0.337389
```

The genome size of *M. genitalium* is 580076 and its GC content $0.157764 + 0.159127 \approx 0.317$.

Answer 285 Generate a random genome:

```
ranseq -s 35 -l 580076 -g 0.317 > ranGenome.fasta
```

and check the result:

```
cchar ranGenome.fasta
# Total number of input characters: 580076
# Char   Count    Fraction
A        197692   0.340804
C        91711    0.158102
G        92097    0.158767
T        198576   0.342328
```

Answer 286 The program `randomizeSeq` shuffles an existing sequence, which keeps its composition unchanged. In contrast, `ranseq` generates a new sequence, the composition of which is bound to vary between runs. You can verify this by piping repeated runs of `randomizeSeq` through `cchar` to find that the result is always the same, while the composition of each run on `ranseq` differs slightly (unless you fix the seed for the random number generator, of course).

Answer 287
$$10 \times 580076 = 5,800,760$$

Answer 288 We need the probability that a nucleotide is not sequenced times the length of the template:
$$580076 \times e^{-10} \approx 26.$$
That is, the expected combined length of all gaps in the assembly is 26.

Answer 289 By solving
$$L \times e^{-c} = 1$$
for c we find
$$c = -\ln\left(\frac{1}{L}\right).$$
In our case, the theoretical coverage $c \approx 13.3$.

Answer 290 Again, we write for the desired coverage
$$c = -\ln\left(\frac{0}{L}\right),$$

but since $\ln(0) = -\infty$, $c = \infty$ in this case. In other words, a combined gap length of 0 cannot be achieved. That is one reason why shotgun sequencing projects usually end with a few gaps that need to be closed by other laboratory methods.

Answer 291 Run `sequencer`:

```
sequencer -s 35 -c 13.3 ranGenome.fasta > reads.fasta
```

Again, we set a seed (`-s`) for the random number generator of `sequencer` to be able to exactly reproduce our result. The number of reads we have just generated is

```
grep -c '^>' reads.fasta
```

```
77161
```

and the number of nucleotides

```
cchar reads.fasta
```

```
# Total number of input characters: 7715050
# Char   Count     Fraction
A        2633810 0.341386
C        1223688 0.158611
G        1223629 0.158603
T        2633923 0.341401
```

Answer 292 Run the hashing program `velveth`:

```
velveth Assem/ 21 -short -fasta reads.fasta
```

Answer 293 First run the assembly on the hashed reads stored in the directory `Assem`:

```
velvetg Assem/ -exp_cov 13.3
```

Then count the contigs

```
grep -c '^>' Assem/contigs.fa
```

```
105
```

Ideally, we would get a single contig, but ending up with multiple contigs is the usual outcome of shotgun sequencing. Next, we count the nucleotides in the contigs:

```
cchar Assem/contigs.fa
```

```
# Total number of input characters: 583465
# Char   Count     Fraction
A        198953    0.340985
C        92269     0.158140
G        92672     0.158830
T        199567    0.342038
c        1         0.000002
g        1         0.000002
t        2         0.000003
```

We get an assembly that is $583, 465 - 580, 076 = 3, 389$ nucleotides longer than the template, the length of which would of course be unknown in a real sequencing experiment. Four of those nucleotides are set in lower case to indicate inferior quality.

Answer 294

```
sequencer -P -s 35 -c 13.3 ranGenome.fasta > reads.fasta
velveth Assem/ 21 -shortPaired -fasta reads.fasta
velvetg Assem/ -exp_cov 13.3 -ins_length 500
grep -c '^>' Assem/contigs.fa
69
cchar Assem/contigs.fa
582931
# Char   Count    Fraction
A        198520   0.340555
C        92144    0.158070
G        92533    0.158737
N        236      0.000405
T        199454   0.342157
a        17       0.000029
c        13       0.000022
g        5        0.000009
t        9        0.000015
```

Paired-end sequencing gives fewer contigs than single-end sequencing even when applied to our idealized random genome. The best possible outcome would be to get a single contig that is identical to the input sequence.

Answer 295 Eliminate the sequencing error

```
sequencer -E 0 -P -s 35 -c 13.3 ranGenome.fasta > reads.fasta
velveth Assem/ 21 -shortPaired -fasta reads.fasta
velvetg Assem/ -exp_cov 13.3 -ins_length 500
grep -c '^>' Assem/contigs.fa
1
cchar Assem/contigs.fa
# Total number of input characters: 580135
# Char   Count    Fraction
A        198554   0.342255
C        92092    0.158742
G        91706    0.158077
N        109      0.000188
T        197672   0.340734
a        1        0.000002
t        1        0.000002
```

Without errors, we get a single contig with very few unknown nucleotides.

Answer 296 Simulate with 1% error:

```
sequencer -E 0.01 -P -s 35 -c 13.3 ranGenome.fasta > reads.
    fasta
velveth Assem/ 21 -shortPaired -fasta reads.fasta
velvetg Assem/ -exp_cov 13.3 -ins_length 500
grep -c '^>' Assem/contigs.fa
```

```
457
cchar Assem/contigs.fa
# Total number of input characters: 607802
# Char   Count    Fraction
A        207343   0.341136
C        96454    0.158693
G        96029    0.157994
N        1361     0.002239
T        206519   0.339780
a        31       0.000051
c        15       0.000025
g        19       0.000031
t        31       0.000051
```

This time there are 457 contigs. Notice also the many $(607, 802 - 580, 076 = 27, 726)$ superfluous nucleotides in our assembly.

Answer 297 To *in silico* shotgun sequence the genome of *M. genitalium* and assemble it, run

```
sequencer -s 35 -c 13.3 mgGenome.fasta > reads.fasta
velveth Assem/ 21 -short -fasta reads.fasta
velvetg Assem/ -exp_cov 13.3
grep -c '^>' Assem/contigs.fa
344
cchar Assem/contigs.fa
# Total number of input characters: 573437
# Char   Count    Fraction
A        197561   0.344521
C        89775    0.156556
G        91005    0.158701
N        10       0.000017
T        195030   0.340107
a        21       0.000037
c        10       0.000017
g        8        0.000014
t        17       0.000030
```

There are 344 contigs.

Answer 298 The midpoint, or median, of $\{2, 2, 3, 4, 5\}$ is 3; the mean is $(2 + 2 + 3 + 4 + 5)/5 = 16/5 = 3.2$.

Answer 299 The total contig length is $2 + 2 + 3 + 4 + 5 = 16$. Now walk along $\mathscr{L}$ from right to left, until the cumulative length covered is at least 8; the element reached then is the N_{50}, in our case 4. The connection to the median is as follows: Rewrite $\mathscr{L}$ as $\mathscr{L}'$ such that each element x is repeated x times:

$$\mathscr{L}' = \{2, 2, 2, 2, 3, 3, 3, 4, 4, 4, 4, 5, 5, 5, 5, 5\}$$

The N_{50} of $\mathscr{L}$ is the median of $\mathscr{L}'$, that is, the average of the 8-th and the 9-th element of $\mathscr{L}'$, which is 4, as expected.

Answer 300 The last line of `velvetg` output is

```
Final graph has 866 nodes and n50 of 25857, max 73815, total
    570552, using 77021/77157 reads
```

that is, $N_{50} = 25857$.

Answer 301 The reverse-sorted contigs are generated using

```
cchar -s Assem/contigs.fa   |
grep '^>'                   |
awk '{print $2}'            |
sort -r -n
```

The median contig length is found by extending the pipeline to look up the midpoint of these sorted lengths:

```
head -n 172 |
tail -n 1
47
```

This is quite different from $N_{50} = 25857$ and illustrates that the relationship between median and N_{50} is indirect, as explained in the Answer to Problem 299.

Answer 302 The program `n50.awk` is

```
{
    s[n++] = $1
    c += $1
}
END{
    while(sum < c/2)
        sum += s[i++]
    print "N_50: " s[i-1]
}
```

The complete pipeline now looks like this:

```
cchar -s Assem/contigs.fa |
grep '^>'                 |
awk '{print $2}'          |
sort -r -n                |
awk -f n50.awk
N_50: 25877
```

Our N_{50} differs from that returned by `velvetg` by $25877 - 25857 = 20$ nucleotides. We do not know why that is the case, but notice that the length quoted in the header of a contig is 20 less than the length of the actual contig:

```
cchar -s Assem/contigs.fa |
head -n 1

>NODE_1_length_18331_cov_10.140527: 18351
```

Answer 303

```
sequencer -P -s 35 -c 13.3 mgGenome.fasta > reads.fasta
velveth Assem/ 21 -shortPaired -fasta reads.fasta
velvetg Assem/ -exp_cov 13.3 -ins_length 500
cchar -s Assem/contigs.fa |
grep '^>'                  |
awk '{print $2}'           |
sort -r -n                 |
awk -f n50.awk
N_50: 81536
```

Paired-end sequencing results in much longer contigs (81536) compared to single-end reads (25857). Do not forget to save this assembly

```
cp Assem/contigs.fa mgAssembly.fasta
```

Answer 304 Here is our generalized suffix tree:

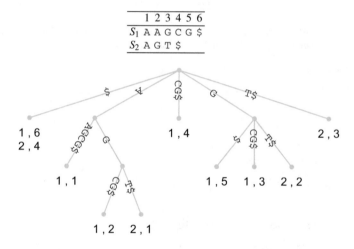

1 2 3 4 5 6
S_1 A A G C G $
S_2 A G T $

Answer 305 Suppose the sequence data is contained in st.fasta. Generate the corresponding generalized suffix tree:

```
drawStrees -i st.fasta -o st
latex st_fig.tex
dvips st_fig.dvi -o
gv st_fig.ps &
```

The result was already shown in the solution to Problem 304.

Answer 306 Mark the string depths (that of the root is always 0):

1 2 3 4 5 6
S_1 A A G C G $
S_2 A G T $

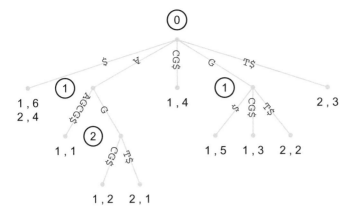

So the longest repeat in the input sequences is AG, which also happens to be the longest repeat *between* S_1 and S_2.

Answer 307

- Length of assembly:

```
cchar mgAssembly.fasta

# Total number of input characters: 577555
# Char   Count     Fraction
A        198323    0.343384
C         89287    0.154595
G         91923    0.159159
N          2774    0.004803
T        194480    0.336730
a           223    0.000386
c           184    0.000319
g           145    0.000251
t           216    0.000374
```

- Length of original sequence:

```
cchar mgGenome.fasta

# Total number of input characters: 580076
# Char   Count     Fraction
A        200544    0.345720
C         91515    0.157764
G         92306    0.159127
T        195711    0.337389
```

The assembly is 2521 nucleotides shorter than the input genome. It also contains 2774 unknown nucleotides (N) and $223 + 184 + 145 + 216 = 768$ low-quality nucleotides shown in lower case.

Answer 308 Generate the starting sequence

```
ranseq -l 1000 > s1.fasta
```

Cut out the first and last 100 bp

```
cutSeq -s -r 1-100,901-1000 s1.fasta > s2.fasta
```

Run mummer

```
mummer s1.fasta s2.fasta | sed '/^#/d'
...
> Rand_1;
        1            1          100
      900          100          101
```

Apart from the messages generated by mummer, the output consists of two lines with entries of the form (x, y, length). Notice that the second hit was extended by one nucleotide; however, solutions will differ as ranseq generates a new sequence every time it is run.

Answer 309 Align and plot

```
mummer s1.fasta s2.fasta |
awk -f mum2plot.awk         |
gnuplot -p mum.gp
```

where mum.gp is

```
set xlabel "s1"
set ylabel "s2"
plot "< cat" title "" with lines
```

to get

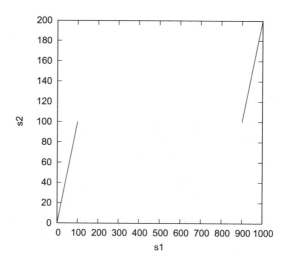

So the sequence in the first file—called reference file in the `mummer` interface—is written along the x-axis, the sequence in the second file—called query file—is written along the y-axis.

Answer 310 Reverse complement `s2.fasta`

```
revComp s2.fasta > s3.fasta
```

and pipe the results of `mummer` into a plot

```
mummer -b -c s1.fasta s3.fasta    |
awk -f mum2plot.awk               |
gnuplot -p mum.gp
```

and plot the `mummer` result

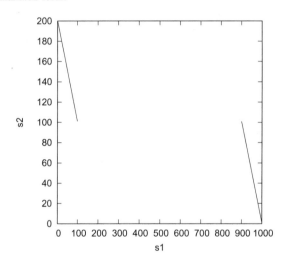

Answer 311 Concatenate `s2.fasta` and `s3.fasta`:

```
cat s2.fasta s3.fasta > s4.fasta
```

Compare the sequences

```
mummer -b -c s1.fasta s4.fasta |
awk -f mum2plot.awk            |
gnuplot -p mum.gp
```

to get

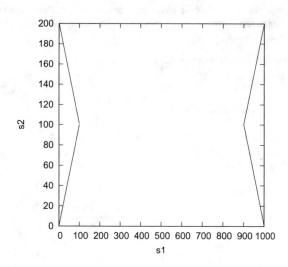

Answer 312 The pipeline

```
mummer -c -b mgGenome.fasta mgAssemblyS.fasta |
awk -f mum2plot.awk                            |
gnuplot -p mg.gp
```

where mg.gp is

```
set xlabel "M. genitalium Reference (kb)"
set ylabel "M. genitalium Assembly (kb)"
plot "< cat" using ($1/1000):($2/1000) title "" with lines
```

gives

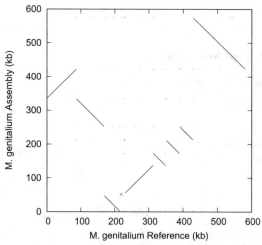

Since we are comparing a known template to its assembly from simulated reads, one might have expected the assembly to be approximately identical to the template. This would have resulted in one line along the main diagonal of the dot plot matrix. However, due to the stochastic nature of shotgun sequencing experiments, all we can expect is that the assembly covers most of the template, which is in fact what we observe. Notice also the transformation of the coordinates to kb, which was achieved with the gnuplot command

```
using ($1/1000):($2/1000)
```

Answer 313 Use the command cchar to find

Strain	Genome length
K12	4,639,675
O157H7	5,528,445

So the genomes of two strains from the same bacterial "species" can differ by $5.5/4.6 \times 100 \approx$ 20% in length and hence in gene content.

Answer 314 The pipeline

```
mummer -c -b ecoliK12.fasta ecoliO157H7.fasta |
awk -f mum2plot.awk                            |
gnuplot -p eco.gp
```

where eco.gp is

```
set xlabel "E. coli K12 (Mb)"
set ylabel "E. coli O157H7 (Mb)"
plot "< cat" using ($1/1000000):($2/1000000) title "" with
    lines
```

gives

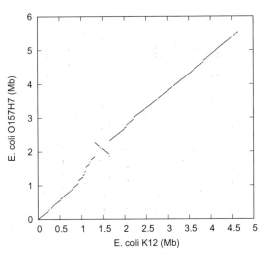

This dot plot shows a large inversion around 1.5 Mb in K12 and 2 Mb in O157H7.

Answer 315 We count the SNPs by counting the entries in the SNP table:

```
tail -n +6 nucmer.snps | wc -l
86321
```

The genome length of $\check{K}12$ is 4,639,675 bp, so the number of pairwise differences per site is

$$\pi = \frac{86,321}{4,639,675} \approx 0.019$$

Answer 316 To compute the divergence time in generations, notice that the pairwise mismatches, π, have accumulated along two diverging lines of descent. Hence, the sought number of generations is

$$g = \frac{0.019/2}{2.2 \times 10^{-10}} \approx 43.2 \times 10^6$$

Answer 317 The lower bound is

$$\frac{g \times 30}{60 \times 24 \times 365.25} \approx 2463 \text{ years,}$$

the upper

$$\frac{g \times 90}{60 \times 24 \times 365.25} \approx 7389 \text{ years.}$$

Answer 318 Sequence the template

```
sequencer -s 10 -c 15 dmChr2L.fasta > reads.fasta
```

Count the reads

```
grep -c '>' reads.fasta
3527065
```

Count the nucleotides sequenced

```
cchar reads.fasta
# Total number of input characters: 352705721
```

and the template length

```
cchar dmChr2L.fasta
# Total number of input characters: 23513712
```

Now calculate the exact coverage

```
bc -l
352705721 / 23513712
15.00000174366344199503
```

This is very close to the expected coverage of 15.

Answer 319 Count the reads shorter than 100 bp:

```
sed '/^>/d' reads.fasta                     |
awk '{if(length($1) < 100)print}'           |
wc -l
12
```

Reads shorter than 100 bp are typically sampled from the edges of the template.

Answer 320 Run

```
bash runBlast.sh > blastTimes.dat
```

where `runBlast.sh` is

```
for a in 1 2 5 10 20 50 100 200 500 1000
do
    echo -n $a
    ((x=$a*2))
    head -n ${x} reads.fasta |
    /usr/bin/time -p blastn -task blastn-short -subject
        dmChr2L.fasta 2>&1 |
    grep real                |
    sed 's/real//'
done
```

Plot `blastTimes.dat`

```
gnuplot -p blastTimes.gp
```

where `blastTimes.gp` is

```
set pointsize 2
set xlabel "Number of Reads"
set ylabel "Run Time (s)"
plot "blastTimes.dat" t "" w linespoints
```

This gives the run time of `blastn` in read mapping mode as a function of the number of reads:

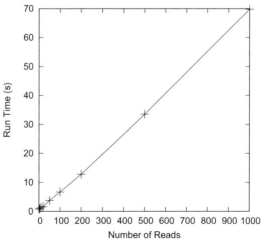

The plot looks linear, so we can extrapolate to estimate the time needed to map all reads:

```
bc -l
3527065 / 1000 * 63.53 / 3600 / 24
2.59345416030092592592
```

In other words, approximately 2 days and 14 h would be needed to align the reads using BLAST in its short-read mode.

Answer 321 Indexing

```
bwa index -p dmChr2L dmChr2L.fasta
```

takes approximately 12 s to complete. The program reports computing the Burrows–Wheeler Transform (BWT) and the suffix array (SA) of the input sequence.

Answer 322 Run the command

```
bash runBwa.sh > bwaTimes.dat
```

where `runBwa.sh` is

```
for a in 1 2 5 10 20 50 100 200 500 1000
do
    echo -n $a
    ((x=$a*2*1000))
    head -n ${x} reads.fasta > tmpReads.fasta
    /usr/bin/time -p bwa mem dmChr2L tmpReads.fasta 2>&1 |
        grep '^real'                                     |
        sed 's/real//'
done
rm tmpReads.fasta
```

Plot `bwaTimes.dat`

```
gnuplot -p bwaTimes.gp
```

where `bwaTimes.gp` is

```
set pointsize 2
set xlabel "Number of Reads (x 1000)"
set ylabel "Run Time (s)"
plot "bwaTimes.dat" t "" w linespoints
```

to get

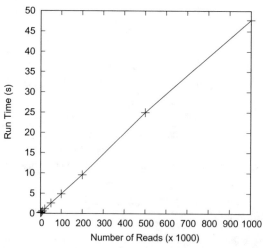

Again, the run time is linear in the number of reads, only roughly a thousand times faster than BLAST. Estimate the time required for mapping all reads

```
bc -l
3527065 / 1000000 * 48.274 / 60
2.83775893016666666666
```

to find that all reads should be mapped in approximately 2 m 50 s.

Answer 323 The command

```
time bwa mem dmChr2L reads.fasta > reads.sam
```

takes 3 m 6 s, a bit more than the estimated 2:50.

Answer 324 Commands like

```
keywordMatcher -p  TGCC... dmChr2L.fasta
>NT_033779.5 Drosophila melanogaster chromosome 2L  TGCC...:
  12300064
```

show `reads.sam` lists every read in its forward orientation.

Answer 325 A read in the forward direction is denoted by dots, in the reverse direction by commas.

Answer 326 Pressing g opens a dialog for entering a position.

Answer 327 To get to the position, press g and enter

```
NT_033779.5:2000
```

As to position annotations, their first digits point to the actual position in the sequence. If you should ever be confused about this, look at the first position:

```
1
C
```

Answer 328 There are 10 proteins and

$$\binom{10}{2} = 45$$

protein pairs, each characterized by up to two edges, so the number of possible edges is 90. Of these 25, that is $25/90 \approx 28\%$, actually exist.

Answer 329 Set up the session

```
mkdir FastLocalAlignmentProt
cd FastLocalAlignmentProt
ln -s ../Data/mgProteome.fasta
```

Count the proteins:

```
grep -c '^>' mgProteome.fasta
476
```

The genome of *M. genitalium* encodes 476 proteins.

Answer 330 An all-against-all comparison is like filling in a square matrix, where query sequences are written along the first column and subject sequences along the top row:

	S_1	S_2	S_3	S_4	S_5
S_1					
S_2					
S_3					
S_4					
S_5					

Every empty cell in that matrix corresponds to a comparison, so there should be $5^2 = 25$ comparisons between our five example sequences, and $476^2 = 226,576$ comparisons between the proteins of *M. genitalium*. To test this prediction, generate five identical sequences

```
for a in $(seq 5); do ranseq -s 13 >> testSeq.fasta; done
```

Run the all-against-all search and count the comparisons

```
blastn -query testSeq.fasta -subject testSeq.fasta |
grep -c 'Score = '
25
```

The result is as predicted.

Answer 331

```
time blastp -max_hsps 1 -outfmt 6 -query mgProteome.fasta
     -subject mgProteome.fasta -evalue 1e-5 > mgProteome.blast
```

This takes 3.5 s. Look at the result:

```
head -n 2 mgProteome.blast
lcl|MG_002 lcl|MG_002 100.000 310   0 0 1 310 1 310 0.0       626
lcl|MG_002 lcl|MG_200  41.667  60 33 1 4  61 9  68 8.90e-09 50.4
```

Answer 332 Constructing the database,

```
time makeblastdb -in mgProteome.fasta -dbtype prot -out
   mgProteome
```

takes 0.02 s. Running the comparisons,

```
time blastp -max_hsps 1 -outfmt 6 -query mgProteome.fasta
     -db mgProteome -evalue 1e-5 > mgProteome.blast
```

takes 3.3 s, about the same as without the database. So, in this case, a precomputed database has no significant effect on the speed of `blastp`.

Answer 333 Carry out the edit and save the result to `tmp`:

```
sed 's/lcl|//g' mgProteome.blast > tmp
```

Make sure the substitution worked

```
head tmp
```

Replace the original

```
mv tmp mgProteome.blast
```

Answer 334 Generate the backup

```
cp mgProteome.blast backup.blast
```

and try

```
sed 's/MG_//g' mgProteome.blast > mgProteome.blast
```

The resulting file is empty. Do not forget to regenerate the original:

```
mv backup.blast mgProteome.blast
```

Answer 335 Print the first ten lines of output

```
head mgProteome.blast
```

to get the tabular BLAST result

q.	s.	% id.	al. len.	mism.	gaps	q. start	q. end	s. start	s. end	eval.	score
002	002	100.000	310	0	0	1	310	1	310	0.0	626
002	200	41.667	60	33	1	4	61	9	68	8.90e-09	50.4
002	019	37.879	66	33	1	4	61	9	74	2.34e-08	48.9
003	003	100.000	650	0	0	1	650	1	650	0.0	1344
003	203	45.426	645	329	9	10	647	5	633	0.0	543
004	004	100.000	836	0	0	1	836	1	836	0.0	1699
004	204	34.366	678	423	10	32	702	20	682	2.11e-125	390
005	005	100.000	417	0	0	1	417	1	417	0.0	851
006	006	100.000	210	0	0	1	210	1	210	1.52e-159	434
007	007	100.000	254	0	0	1	254	1	254	0.0	509

where we have removed the MG_ in front of each accession to make the table fit the page. The first nonself pair is MG_002 and MG_200. Look at the header of MG_002

```
grep MG_002 mgProteome.fasta
>lcl|MG_002 DnaJ domain protein
```

and of MG_200

```
grep MG_200 mgProteome.fasta
>lcl|MG_200 DnaJ domain protein
```

to find they are both DnaJ domain proteins. These belong to the group of molecular chaperones involved in protein folding and cellular stress response.

Answer 336

```
wc -l mgProteome.blast
836 mgProteome.blast
```

Reading a table containing 836 entries might become rather tedious...

Answer 337 The pipeline we are looking for is

```
cut -f 1 mgProteome.blast |
sort                      |
uniq -c                   |
sort -n -r                |
head
```

which gives

```
        17 MG_410
        17 MG_180
        17 MG_179
        16 MG_526
        16 MG_467
        16 MG_303
        16 MG_290
        16 MG_187
        16 MG_065
        16 MG_042
```

In other words, MG_410, MG_180, and MG_179 each have 17 hits in the proteome. Since one of these is a self-hit, they all have at least 16 homologues.

Answer 338 Extract the seventeen proteins linked to MG_410:

```
grep MG_410 mgProteome.blast | # Extract hits to MG_410
cut -f 2                      | # Cut subject column
sort                          |
uniq > protFam.txt
```

and write the protein family to one line

```
tr '\n' ' ' < protFam.txt
```

to get

```
MG_014 MG_015 MG_042 MG_065 MG_079 MG_080 MG_119 MG_179 MG_180
MG_187 MG_290 MG_303 MG_304 MG_410 MG_421 MG_467 MG_526
```

The conversion from single column to single row makes printing easier.

Answer 339 Execute

```
neato -T x11 example2.dot
```

where example2.dot is

```
graph G {
        1 -- 2
        2 -- 3
        2 -- 5
        4 -- 5
}
```

to get the graphic depicted in the problem.

Answer 340 Here is the dot code for specifying the figure:

```
graph G {
        2 -- 1 [dir=forward]
        2 -- 5 [dir=forward]
        3 -- 4 [dir=forward]
        4 -- 1 [dir=both]
        5 -- 1 [dir=forward]
        5 -- 3 [dir=forward]
}
```

Answer 341 The expected number of edges is the maximal number of edges times the edge probability:

$$10 \times 9 \times 0.5 = 45.$$

Take a look at the output file

```
cat ranDot.dot
graph G {
        1 -- 3 [dir=forward]
        1 -- 5 [dir=forward]
        1 -- 6 [dir=both]
        1 -- 7 [dir=both]
...
```

We can thus compute the number of observed edges by filtering for the two types of edges, "forward" and "both", and counting "both" twice, "forward" once:

```
awk '/both/{s+=2}/forward/{s++}END{print s}' ranDot.dot
43
```

Your result may well differ from ours, but it should also be close to the expectation (45).

Answer 342 First apply `neato`

```
neato -T x11 ranDot.dot
```

to get

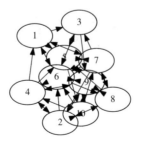

This is rather messy. Try `circo`

```
circo -T x11 ranDot.dot
```

to get

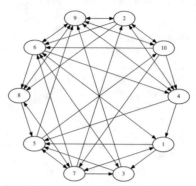

which is a nicer layout for a highly connected graph such as ours.

Answer 343 *M. genitalium* contains one highly connected protein family. It is much better resolved by `circo` than `neato`; hence, we used

```
circo -T x11 mgProteome.dot
```

to get

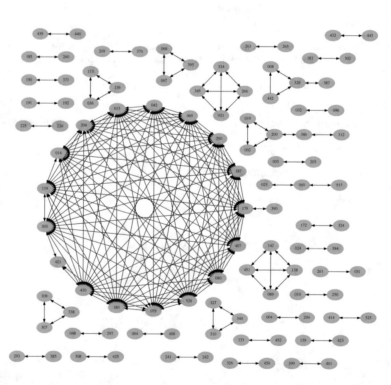

The protein family at the center contains the seventeen proteins already in `protFam.txt` plus MG_390. To record this, write

```
cp protFam.txt protFam2.txt
echo 'MG_390' >> protFam2.txt
```

where `protFam2.txt` is now

```
tr '\n' ' ' < protFam2.txt
MG_014 MG_015 MG_042 MG_065 MG_079 MG_080 MG_119 MG_179 MG_180
MG_187 MG_290 MG_303 MG_304 MG_410 MG_421 MG_467 MG_526 MG_390
```

Answer 344 Calculate the new layout

```
circo -T x11 mgProteome2.dot
```

to get

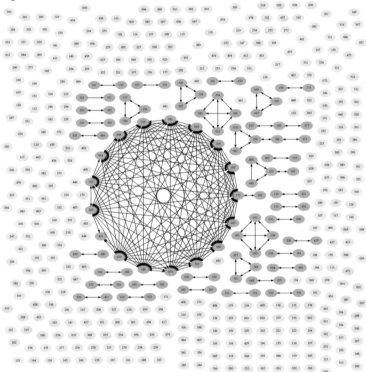

Compute the number of singletons:

```
awk 'NF==1{if(length($1)==3)c++}END{print c}' mgProteome2.dot
374
```

That is, $374/476 \times 100 \approx 78.6\%$ of the proteome consists of singletons, given $E \leq 10^{-5}$.

Answer 345 Construct the list of alternatives matches

```
tr '\n' '|' < protFam2.txt
```

and then copy and paste them into the command

```
grep -E '(MG_014|MG_015|...|MG_390)' mgProteome.fasta
```

to get

```
>lcl|MG_014 ABC transporter, ATP-binding
>lcl|MG_015 ABC transporter, ATP-binding
>lcl|MG_042 spermidine
>lcl|MG_065 ABC transporter, ATP-binding protein
>lcl|MG_079 oligopeptide ABC transporter, ATP-binding protein
>lcl|MG_080 oligopeptide ABC transporter, ATP-binding protein
>lcl|MG_119 ABC transporter, ATP-binding protein
>lcl|MG_179 metal ion ABC transporter, ATP-binding protein, putative
>lcl|MG_180 metal ion ABC transporter ATP-binding protein, putative
>lcl|MG_187 ABC transporter, ATP-binding protein
>lcl|MG_290 phosphonate ABC transporter, ATP-binding protein, putative
>lcl|MG_303 metal ion ABC transporter, ATP-binding protein, putative
>lcl|MG_304 metal ion ABC transporter, ATP-binding protein, putative
>lcl|MG_390 ABC transporter, ATP-binding
>lcl|MG_410 phosphate ABC transporter, ATP-binding protein
>lcl|MG_421 excinuclease ABC, A subunit
>lcl|MG_467 ABC transporter, ATP-binding protein
>lcl|MG_526 ABC transporter, ATP-binding protein
```

Most of these are annotated as ABC transporters, so our protein family consists of ABC transporters.

Answer 346 Open `prosite.doc` in emacs or `less` and search for *ABC transporter*. Here is an edited version of the relevant entry:

```
{PDOC00185}
{PS00211; ABC_TRANSPORTER_1}
{PS50893; ABC_TRANSPORTER_2}
{BEGIN}
*********************************************************************
* ATP-binding cassette, ABC transporter-type, signature and profile *
*********************************************************************

ABC  transporters  belong  to the ATP-Binding Cassette (ABC) superfamily
which uses the hydrolysis of ATP to energize diverse biological systems.
ABC transporters are  minimally constituted  of  two conserved regions:
a highly conserved  ATP  binding  cassette  (ABC)  and  a less conserved
transmembrane  domain (TMD). These regions can  be  found  on  the  same
protein or on two different ones.  Most  ABC  transporters function as a
dimer and therefore are constituted of four domains, two ABC modules and
two TMDs [1].

ABC  transporters are involved in the export or import of a wide variety
of  substrates  ranging  from  small  ions  to macromolecules. The major
function of  ABC  import  systems  is  to provide essential nutrients to
bacteria. They are found only in prokaryotes and their four constitutive
domains  are  usually  encoded  by  independent  polypeptides (two  ABC
proteins and two TMD proteins). Prokaryotic importers require additional
extracytoplasmic binding proteins (one or more per systems) for function.
In  contrast, export systems are  involved  in  the extrusion of noxious
substances, the export  of  extracellular  toxins  and  the targeting of
membrane components. They  are  found  in  all  living organisms and  in
general  the TMD is fused to the ABC module in a variety of combinations.
Some  eukaryotic  exporters  encode  the  four  domains on the  same
polypeptide chain [2,3].
...
{END}
```

Answer 347 A lot of energy-consuming transfer of various molecules across the bacterial cell membrane is conducted via ABC transport proteins.

Answer 348 Set up the working directory

```
mkdir PsiBlast
cd PsiBlast
ln ../Data/mgProteome.fasta
```

Convert the proteome of *M. genitalium* to a BLAST database

```
makeblastdb -dbtype prot -in mgProteome.fasta -out mgProteome
```

Get the sequence of protein M_410

```
getSeq -s 410 mgProteome.fasta > mg_410.fasta
```

and compare it to the proteome using `psiblast`:

```
psiblast -query mg_410.fasta -db mgProteome -outfmt 6 >
    mg_410.psi
```

The syntax should look familiar from our previous work with `blastn` and `blastp`. Count the number of unique hits with $E \leq 10^{-5}$:

```
awk '{if($11<=10^-5)print $1 "\t" $2}' mg_410.psi       |
tr '\t' '\n'                                            |
sort                                                    |
uniq                                                    |
wc -l
17
```

We found 17 members of this protein family.

Answer 349 Run `blastp`:

```
blastp -query mg_410.fasta -db mgProteome -outfmt 6 > mg_410.bp
```

Visual inspection of `mg_410.psi` and `mg_410.bp` shows that they are at least highly similar. To confirm identity, use

```
diff mg_410.psi mg_410.bp
```

Answer 350 Run `psiblast` iteratively

```
psiblast -num_iterations 0 -query mg_410.fasta -db mgProteome
    -outfmt 6 > mg_410b.psi
```

To count the number of rounds, look for the number of comparisons with MG_410 as query and subject

```
awk '{if($1~/410/ && $2~/410/)c++}END{print c}' mg_410b.psi
4
```

Now find the row numbers where each round starts

```
awk '{if($1~/410/ && $2~/410/)print NR}' mg_410b.psi
1
49
104
155
```

Finally, count the number of distinct hits with $E \leq 10^{-5}$ from line 155 onward:

```
tail -n +155 mg_410b.psi                    | # Print line >= 155
grep MG                                     | # Filter out footer
awk '{if($11<=10^-5)print $1 "\n" $2}'      | # Check E-value
sort                                        |
uniq                                        |
wc -l                                       |
21
```

Answer 351 Get `protFam2.txt`:

```
cp ../FastLocalAlignmentProt/protFam2.txt .
```

and look for the differences between the lists

```
diff psiBlastList.txt protFam2.txt
7,8d6
< MG_107
< MG_110
14d11
< MG_298
17d13
< MG_390
21a18
> MG_390
```

Three extra proteins MG_107, MG_110, and MG_298 were found by `psiblast`, while MG_390 is contained in both lists, but at different positions.

Answer 352 Use `grep` with extended notation to extract the header lines of the three extra proteins:

```
grep -E '(MG_107|MG_110|MG_298)' mgProteome.fasta
>lcl|MG_107 guanylate kinase
>lcl|MG_110 ribosome small subunit-dependent GTPase A
>lcl|MG_298 chromosome segregation protein SMC
```

These annotations are not in any obvious way connected to "ABC transporter". However, the kinase binds ATP, the GTPase A binds GTP, which is similar to ATP, and SMC proteins belong to the ATPases. That is to say, ATP binding is the common feature of the 21 proteins we have identified.

Answer 353 Run `psiblast`

```
psiblast -out_ascii_pssm psiBlast.mat -num_iterations 0 \
-query mg_410.fasta -db mgProteome -outfmt 6 > mg_410b.psi
```

The position-specific score matrix is contained in the first 20 columns labeled A for alanine through V for valine. It consists of 329 rows, one for each amino acid in MG_410.

Answer 354 Identify the most frequent amino acid

```
cchar mg_410.fasta |
sed '/^#/d'         |
sort -k 2 -n -r     |
head -n 1
I    41   0.124620
```

Extract all positions occupied by isoleucine:

```
awk 'NR==3 || $2=="I"{print}' psiBlast.mat
```

which returns the position-specific score information

```
      A  R  N  D  C  Q  E  G  H  I  L  K  M  F  P  S  T  W  Y  V ...
 5 I -1 -3 -3 -3 -1 -3 -3 -4 -3  4  1 -3  1  0 -3 -2 -1 -3 -1  2 ...
19 I -1 -1 -1  0 -2  0  3 -3 -1  2  0 -1  0 -1 -2 -1 -1 -3 -2  1 ...
30 I -2 -3 -4 -4 -1 -3 -3 -4 -3  3  4 -3  2  0 -3 -3 -1 -2 -1  2 ...
...
```

The match score is the score for isoleucine (I); to find its range, extend the previous command

```
awk 'NR==3 || $2=="I"{print}' psiBlast.mat       |
tail -n +2                                        |
awk '{print $12}'                                 |
sort                                              |
uniq                                              |
tr '\n' ' '
0 1 2 3 4 5
```

The match score for isoleucine ranges between 0 and 5. The corresponding BLOSUM62 score is 4.

Answer 355 The first three accessions need to be extended to achieve uniqueness:

```
getSeq -s 'HBA_HUMAN H' uniprot_sprot.fasta > hbaHuman.fasta
getSeq -s 'HBA_HORSE H' uniprot_sprot.fasta > hbaHorse.fasta
getSeq -s 'HBB_HUMAN H' uniprot_sprot.fasta > hbbHuman.fasta
getSeq -s HBB_HORSE     uniprot_sprot.fasta > hbbHorse.fasta
```

Alternatively, `getSeq` could have been applied twice, for example

```
getSeq -s HBA_HUMAN uniprot_sprot.fasta | getSeq -s Hem
```

Answer 356 Construct the file containing the subject sequences:

```
cat hbbHuman.fasta hbaHorse.fasta hbbHorse.fasta >
  subject.fasta
```

Then run `blastp`

```
blastp -query hbaHuman.fasta -subject subject.fasta -outfmt 2
```

to get the slightly edited query-anchored alignment

```
Q_1   1   MVLSPADKTNVKAAWGKVGAHAGEYGAEALERMFLSFPTTKTYFPHFDLSHGSAQVKGHG   60
S_2   1   MVLSAADKTNVKAAWSKVGGHAGEYGAEALERMFLGFPTTKTYFPHFDLSHGSAQVKAHG   60
S_1   4    LTPEEKSAVTALWGKV--NVDEVGGEALGRLLVVYPWTQRFFESFDLSMGNPKVKAHG    65
                                      \     \
                                      |     |
                                      G  TPDAV
S_3   3    LSGEEKAAVLALWDKVNEE--EVGGEALGRLLVVYPWTQRFFDSFDLSNGNPKVKAHG    64
                                      \     \
                                      |     |
                                      G  PGAVM


Q_1   61  KKVADALTNAVAHVDDMPNALSALSDLHAHKLRVDPVNFKLLSHCLLVTLAAHLPAEFTP  120
S_2   61  KKVGDALTLAVGHLDDLPGALSNLSDLHAHKLRVDPVNFKLLSHCLLSTLAVHLPNDFTP  120
S_1   66  KKVLGAFSDGLAHLDNLKGTFATLSELHCDKLHVDPENFRLLGNVLVCVLAHHFGKEFTP  125
S_3   65  KKVLHSFGEGVHHLDNLKGTFAALSELHCDKLHVDPENFRLLGNVLVVVLARHFGKDFTP  124


Q_1  121  AVHASLDKFLASVSTVLTSKYR   142
S_2  121  AVHASLDKFLSSVSTVLTSKYR   142
S_1  126  PVQAAYQKVVAGVANALAHKY   146
S_3  125  ELQASYQKVVAGVANALAHKY   145
```

The amino acids printed below

```
\
|
```

in Subjects `S_1` and `S_3` are insertions.

Answer 357 Construct another set of subject sequences:

```
cat hbaHuman.fasta hbaHorse.fasta hbbHorse.fasta >
   subject2.fasta
```

and run `blastp`

```
blastp -query hbbHuman.fasta -subject subject2.fasta -outfmt 2
```

to get the query-anchored alignment

```
Q_1  2   VHLTPEEKSAVTALWGKVNVDEVGGEALGRLLVVYPWTQRFFESFGDLSTPDAVMGNPKV  61
S_3  1   VQLSGEEKAAVLALWDKVNEEEVGGEALGRLLVVYPWTQRFFDSFGDLSNPGAVMGNPKV  60
S_1  3    LSPADKTNVKAAWGKVHAGEYGAEALERMFLSFPTTKTYFPHF-DLS-----HGSAQV   56
                             \
                             |
                             GA
S_2  3    LSAADKTNVKAAWSKVHAGEYGAEALERMFLGFPTTKTYFPHF-DLS-----HGSAQV   56
                             \
                             |
                             GG


Q_1  62  KAHGKKVLGAFSDGLAHLDNLKGTFATLSELHCDKLHVDPENFRLLGNVLVCVLAHHFGK  121
S_3  61  KAHGKKVLHSFGEGVHHLDNLKGTFAALSELHCDKLHVDPENFRLLGNVLVVVLARHFGK  120
S_1  57  KGHGKKVADALTNAVAHVDDMPNALSALSDLHAHKLRVDPVNFKLLSHCLLVTLAAHLPA  116
S_2  57  KAHGKKVGDALTLAVGHLDDLPGALSNLSDLHAHKLRVDPVNFKLLSHCLLSTLAVHLPN  116


Q_1  122  EFTPPVQAAYQKVVAGVANALAHKYH  147
S_3  121  DFTPELQASYQKVVAGVANALAHKYH  146
S_1  117  EFTPAVHASLDKFLASVSTVLTSKY   141
S_2  117  DFTPAVHASLDKFLSSVSTVLTSKY   141
```

Now, there is one insertion of two amino acids, `GA` into HBA_HUMAN and `GG` into HBA_HORSE.

Answer 358 Use commands like

```
gal -p -i hbaHuman.fasta -j hbaHorse.fasta -m BLOSUM62 |
grep Score
```

to get

	HBA_HUMAN	HBA_HORSE	HBB_HUMAN	HBB_HORSE
HBA_HUMAN	—	648	282	268
HBA_HORSE	0.00	—	264	266
HBB_HUMAN	0.56	0.59	—	633
HBB_HORSE	0.59	0.59	0.02	—

where scores are in the top triangle and distances in the bottom triangle.

Answer 359 The most similar sequences, HBA from human and horse, are clustered first, followed by the HBBs; finally, the two groups are merged to give the following guide tree:

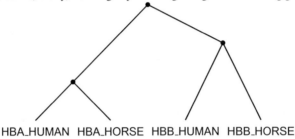

HBA_HUMAN   HBA_HORSE   HBB_HUMAN   HBB_HORSE

Answer 360 The pairs α/α and β/β are orthologs marked by solid lines, the α/β pairs are paralogs marked by dotted lines:

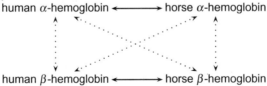

Answer 361 Run `clustalw`

```
clustalw hemoglobin.fasta
```

A slightly edited version of the guide tree in `hemoglobin.dnd` looks like this

```
(
(
HBA_HORSE:0.06,
HBA_HUMAN:0.06)
:0.4,
HBB_HORSE:0.08,
HBB_HUMAN:0.08);
```

Its graphical representation is

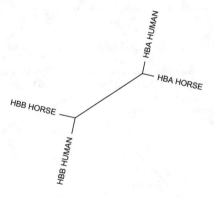

Answer 362 There are four gaps. The left-most has two origins,

```
-V
MV
```

was introduced when aligning the two β-hemoglobins, while

```
-M
-M
```

was introduced when the α- and β-pairs were aligned. The other three gaps each affect a pair of sequences, so they must have been introduced when the α- and β-pairs were aligned.

Answer 363 The `clustalw` alignment generates an end gap:

```
S1                ATG
S2                -AG
                   *
```

Answer 364 Values ≤ 5 give the alternative gap pattern, for example,

```
clustalw hemoglobin.fasta -GAPOPEN=5
```

Answer 365 Run the command

```
bash simTimes1.sh > simTimes1.dat
```

where `simTimes1.sh` is

```
for a in 100 200 500 1000 2000 5000 10000 20000
do
    echo -n ${a} ' '
    ranseq -n 2 -l ${a} > test.fasta
    /usr/bin/time -p clustalw test.fasta 2>&1      |
        grep real                                  |
        sed 's/real //'
done
rm test.fasta
```

to collect the times. Plot them

```
gnuplot -p simTimes1.gp
```

where `simTimes1.gp` is

```
set xlabel "Sequence Length (kb)"
set ylabel "Time (s)"
plot "simTimes1.dat" using ($1/1000):2 title "" with
    linespoints
```

to get the run time of `clustalw` as a function of sequence length:

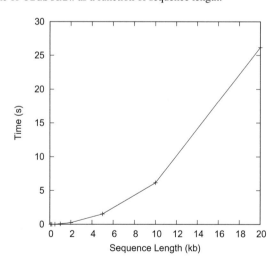

As the sequence length is doubled, the run time is roughly quadrupled.

Answer 366 Run

```
bash simTimes2.sh > simTimes2.dat
```

to collect run times, where `simTimes2.sh` is

```
for a in 2 5 10 20 50 100
do
    echo -n ${a} ' '
    ranseq -n ${a} -l 1000 > test.fasta
    /usr/bin/time -p clustalw test.fasta 2>&1        |
        grep real                                    |
        sed 's/real //'
done
```

Plot them

```
gnuplot -p simTimes2.gp
```

where `simTimes2.gp` is

```
set xlabel "Number of Sequences"
set ylabel "Time (s)"
plot "simTimes2.dat" title "" with linespoints
```

to get

Again, as the number of sequences doubles, the run time roughly quadruples.

Answer 367 Get the sample

```
getSeq -s Hemoglobin uniprot_sprot.fasta    |
awk -f fasta2tab.awk                         |
awk -f shuffle.awk                           |
head -n 100                                  |
tr '\t' '\n'                                 |
fold > hb100.fasta
```

Make sure 100 sequences were obtained

```
grep -c '^>' hb100.fasta
100
```

Measure the run time of `clustalw`

```
/usr/bin/time -p clustalw hb100.fasta 2>&1 |
grep real
```

which takes 2.30 s. Aligning 200 hemoglobin sequences takes 8.92 s, roughly four times longer.

Answer 368 There are

```
getSeq -s Hemoglobin uniprot_sprot.fasta | grep -c '^>'
833
```

hemoglobin sequences in UniProt. Aligning all of them with `clustalw` would roughly take $2.3 \times 4^3 = 147s$.

Answer 369 Get the hemoglobin sequence:

```
getSeq -s Hemoglobin uniprot_sprot.fasta > hemoglobinAll.fasta
```

Count them, just to make sure:

```
grep -c '^>' hemoglobinAll.fasta
833
```

Align them

```
time clustalw hemoglobinAll.fasta > /dev/null
```

which takes 150.1 s. This is close to the predicted run time of 147 s.

Answer 370 Set up the session:

```
mkdir TreesOfLife
cd TreesOfLife
```

Without a scale bar the absolute branch lengths are meaningless, but since the branches all look the same, we can write

```
((A:1,B:1):1,(C:1,D:1):1);
```

Answer 371 The command

```
new2view first.tree
```

yields

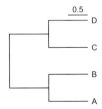

Your scaling might differ from ours—this can be adjusted with the -d and -s options. More-over, the tree is now drawn from left to right rather than top to bottom. Left to right is often used when drawing phylogenies, as this makes it easier to place the taxon labels. However, when talking about the children of a node, we shall continue to refer to the right and the left child, rather than the top and the bottom child.

Answer 372 `second.tree` contains

`((A,B),(C,D));`

which looks identical to the tree with unit branch lengths—but that is just the drawing convention adopted by `new2view`:

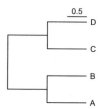

Answer 373 The Newick tree is now

`((,),(,))root;`

and looks like

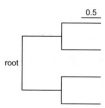

Answer 374 In Newick notation, we have

`((B,A),(D,C));`

which is rendered with switched leaf labels compared to `second.tree`:

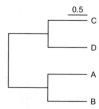

However, in phylogenies only the branching pattern matters, and hence a phylogeny remains unchanged by rotating branches around nodes.

Answer 375 Visualizing all tree files at once using

```
new2view -u *.tree
```

yields

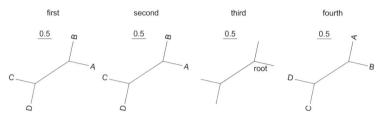

An unrooted tree does not have a unique starting point. It still has direction, though, going from internal to leaf nodes. The unrooted layout is also known as "radial".

Answer 376 Save the four trees

```
cat first.tree second.tree third.tree fourth.tree >
    fourTreesU.tree
```

edit them to obtain

```
(A:1,B:1,(C:1,D:1):1);
(A,B,(C,D));
(,,(,))root;
(B,A,(D,C));
```

and draw them

```
new2view fourTreesU.tree
```

to get the expected radial layout—notice the root position in the third tree. This radial layout is the standard representation of unrooted trees in biology.

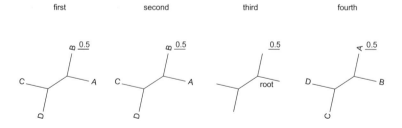

When enforcing the rooted layout

```
new2view -r fourTreesU.tree
```

we see the trifurcating root:

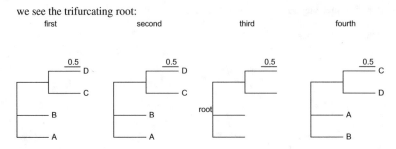

Answer 377 The preorder sequence is 2, 1, 6, 4, 3, 5, 8, 7, and 9.

Answer 378 The inorder sequence is 1, 2, 3, 4, 5, 6, 7, 8, and 9.

Answer 379 The postorder sequence is 1, 3, 5, 4, 7, 9, 8, 6, and 2.

Answer 380 The commands for inorder, preorder, and postorder traversal are

```
traverseTree -t inorder traverse.tree
traverseTree -t preorder traverse.tree
traverseTree -t postorder traverse.tree
```

where `traverse.tree` contains

```
(,((,),(,)));
```

All three methods are centered on visiting child nodes rather than neighbor nodes. Hence, the name "depth first" traversal.

Answer 381 The command

```
genTree | new2view
```

gives, for example

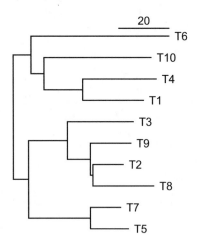

Your tree is bound to differ, but its leaves will also not be perfectly lined up. The leaves are labeled T1,..., Tn for *taxon* 1,...,n. With default settings, `genTree` produces branch lengths proportional to the number of mutations along a given branch. Since this is a random variable, branches rarely end at the same point along the horizontal axis. This interpretation of branch lengths as mutations seems to conflict with our intuition that time is marked along the x-axis: present on the right, past on the left. The conflict is resolved when we realize that the number of mutations is drawn from a Poisson distribution with mean proportional to the time that has elapsed between a given node and its parent. If we knew the time, the leaves would all align.

Answer 382 Repeat

```
genTree -t 0 | new2view
```

twice to get, for example

T7	T9
T2	T3
T3	T1
T8	T5
T1	T4
T10	T10
T5	T8
T4	T2
T9	T7
T6	T6

The trees differ in the order of taxa.

Answer 383 Compute $n!$ as a function of n:

```
awk -f factorial.awk -v n=100 > factorial.dat
```

where `factorial.awk` is

```
BEGIN{
    if(!n)    # n set via -v?
        n = 100
    f = 1
    for(i=1; i<=n; i++){
        f *= i
        print i, f
    }
}
```

Plot the results

```
gnuplot -p plot.gp
```

where `plot.gp` is

```
set xlabel "n"
set ylabel "n!"
set logscale y
plot[][] "factorial.dat" t "" w l
```

to get

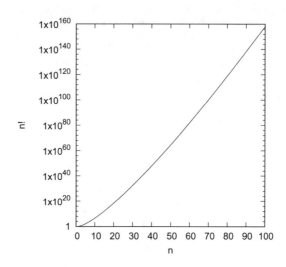

This gives a first impression of how the number of phylogenies scale with n.

Answer 384 Compute the number of trees

```
awk -f numTrees.awk -v n=100 > numTrees.dat
```

where `numTrees.awk` is

```
BEGIN{
    if(!n)   # n set via -v?
        n = 100
    f = 1
    if(n>0)
        print 1, f
    if(n>1)
        print 2, f
    for(i=3; i<=n; i++){
        f = 1
        for(j=3; j<=(2*i-3); j+=2)
            f *= j
        print i, f
    }
}
```

Plot the result together with n!

```
gnuplot -p plot2.gp
```

where `plot2.gp` is

```
set xlabel "n"
set logscale y
plot[][] "numTrees.dat" t "Number of Trees" w l,\
"factorial.dat" t "n!" w l
```

to get the exact number of phylogenies as a function of sample size, n, compared to the estimation via $n!$. The factorial estimation is roughly 20 orders of magnitude too small when $n = 100$:

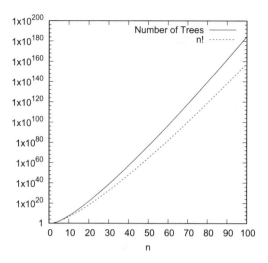

Answer 385 Use the command

```
genTree -s -t 0 | new2view
```

to get something similar to

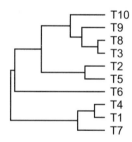

This time all the leaves line up because rather than representing mutations, the branches go back to the simulated times of species divergence.

Answer 386 Use, for example,

```
genTree -s -S 13
```

and

```
genTree -S 13
```

to get

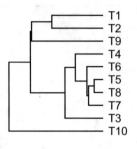
Standard Clock

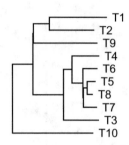
Molecular Clock

This illustrates that the molecular clock is a stochastic clock that behaves only approximately like its standard version.

Answer 387 Here is the corresponding trace of the UPGMA algorithm:

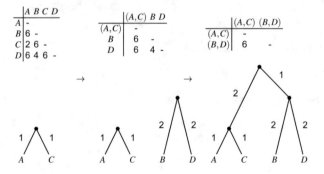

Answer 388 The distances

- fulfill the three point criterion,
- fit the UPGMA tree,
- and, by implication, for n taxa there are no more than $n - 1$ distinct entries in the distance matrix;

in other words, they are ultrametric.

Answer 389 Get the headers of the *Hominidae* sequences:

```
grep '^>' hominidae.fasta

>Pongo
>Gorilla
>Homo
>Pan
```

Answer 390 Here are the full scientific names and the trivial names of the *Hominidae* in
`hominidae.fasta`:

Name in File	Full Name	Trivial Name
Homo	*Homo sapiens*	human
Pan	*Pan troglodytes/paniscus*	chimp
Gorilla	*Gorilla gorilla*	gorilla
Pongo	*Pongo pygmaeus*	orangutan

Answer 391 Extract the first ten polymorphic positions:

```
gd -P hominidae.fasta        |  # get polymorphisms
getSeq -c -s Pos             |  # exclude positions
cutSeq -r 1-10

>Homo 1..10
GTCATTCACC
>Pan 1..10
ATTATCCACC
>Gorilla 1..10
GTTGTTTATC
>Pongo 1..10
ACCACCCGTT
```

to compute the distance matrix

	Homo	*Pan*	*Gorilla*	*Pongo*
Homo	-			
Pan	3	-		
Gorilla	4	5	-	
Pongo	7	6	9	-

Answer 392 The file `test.dist` should look like this:

```
cat test.dist

4
Ho 0 3 4 7
Pa 3 0 5 6
Go 4 5 0 9
Po 7 6 9 0
```

We cluster the taxa in `test.dist` by tracing the UPGMA algorithm:

```
clustDist -u -m test.dist
****** Round 1 ******
Ho        0    3.00    4.00    7.00
Pa     3.00       0    5.00    6.00
Go     4.00    5.00       0    9.00
Po     7.00    6.00    9.00       0
****** Round 2 ******
Go           0    9.00    4.50
Po        9.00       0    6.50
Ho,Pa     4.50    6.50       0
****** Round 3 ******
Po           0    7.75
Go,Ho,Pa  7.75       0
(Po:3.875,(Go:2.250,(Ho:1.500,Pa:1.500):0.750):1.625);
```

To visualize the tree, run

```
clustDist -u test.dist | new2view -d 3 -s 0.7
```

where −d 3 restricts the smallest dimension of the tree to 3 cm, and −s 0.7 sets the length of the scale bar, to get

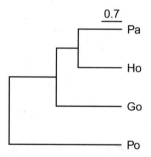

Answer 393 Compute the distances

```
dnaDist hominidae.fasta > hominidae.dist
```

and print them to the screen:

```
cat hominidae.dist
4
Homo       0.000000 0.093798 0.111717 0.180872
Pan        0.093798 0.000000 0.113013 0.192322
Gorilla    0.111717 0.113013 0.000000 0.188008
Pongo      0.180872 0.192322 0.188008 0.000000
```

Since this matrix has more than three distinct entries, the distances are not ultrametric.

Answer 394 The command

```
clustDist -u hominidae.dist |
new2view
```

gives the *Hominidae* phylogeny

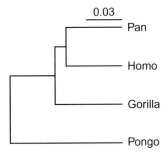

Answer 395 Count the primate taxa contained in `primates.fasta`

```
grep -c '^>' primates.fasta
27
```

Compute the sequence lengths of their mitochondrial genomes and sort them

```
cchar -s primates.fasta    |
grep '^>'                  |
sort -k 2
```

to find they range between 15467 and 17036 bp.

Answer 396 Our measurements of the "real" time were as follows:

Threads	Time (s)
1	2.087
2	1.093
4	0.611
8	0.407

Your exact measurements are bound to differ slightly from ours, but the trend should be similar. If a lot of computing is going on while you are making these measurements, eight threads would not result in as much speedup as we observed when making these runs on an otherwise idle machine.

Answer 397 Cluster the distances and draw the primate phylogeny:

```
andi -t 8 primates.fasta   |
clustDist -u               |
new2view -d 12
```

which gives

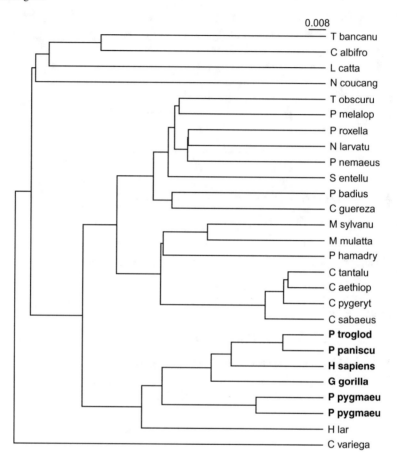

The branching order for the *Hominidae* in **bold** is the same as with the small data set.

Answer 398 By plugging the distances into the equation describing the four point criterion, we get

$$7 + 7 = 10 + 4 \geq 5 + 5,$$

which is true.

Answer 399 In Fig. 5.4a, we have $d_{AB} = 5$, $d_{AC} = 7$, and $d_{BC} = 4$. Since these are three distinct numbers, the three point criterion does not hold.

Answer 400 Our trace of neighbor-joining begins with the row sums:

	A	B	C	D	r_i
A	-	5	7	10	**22**
B		-	4	7	**16**
C			-	5	**16**
D				-	**22**

Answer 401

	A	B	C	D	r_i
A	-	5	7	10	22
B	**-14**	-	4	7	16
C	**-12**	**-12**	-	5	16
D	**-12**	**-12**	**-14**	-	22

Answer 402

	C	D	(A, B)
C	-	5	3
D		-	6
(A, B)			-

Answer 403

$$d_{A(AB)} = (2 \times 5 + 22 - 16)/4 = 4,$$

and

$$d_{B(AB)} = (2 \times 5 + 16 - 22)/4 = 1.$$

Answer 404

$$d_{rC} = (5 + 3 - 6)/2 = 1$$
$$d_{rD} = (5 + 6 - 3)/2 = 4$$
$$d_{r(AB)} = (3 + 6 - 5)/2 = 2$$

Answer 405 Set up the session

```
mkdir NeighborJoining
cd NeighborJoining
```

First, trace the clustering procedure

```
clustDist -m test.dist
****** Round 1 ******
A        0    5.00    7.00  10.00 22.00
B     0.00       0    4.00   7.00 16.00
C     0.00    0.00       0   5.00 16.00
D     0.00    0.00    0.00      0 22.00
****** Round 2 ******
C        0    5.00    3.00
D     5.00       0    6.00
A,B   3.00    6.00       0
(C:1.000000,D:4.000000,(A:4.000000,B:1.000000):2.000000);
```

Then save the unrooted tree

```
clustDist test.dist > testU.tree
```

and draw it

```
new2view testU.tree
```

to get

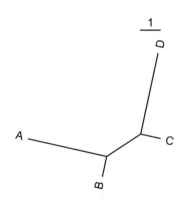

Answer 406 Execute

```
new2view -d 3 testR.tree
```

where testR.tree is the midpoint-rooted phylogeny

```
((C:1,D:4):1,(A:4,B:1):1);
```

to get

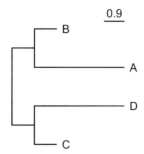

Answer 407 Use `testU.tree` as input to `retree`. This is started by entering

```
phylip retree
```

or just

```
retree
```

depending on your setup. Then follow the menu. PHYLIP writes the new tree to the file `outtree`. After quitting PHYLIP, execute

```
new2view -d 3 outtree
```

where `outtree` is

```
cat outtree
((A:4.0,B:1.0):1.0,(C:1.0,D:4.0):1.0);
```

to get

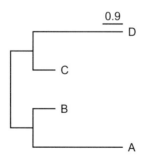

which has the same topology as the result computed by hand.

Answer 408 Compute the neighbor-joining tree

```
andi -t 8 primates.fasta |
clustDist > primates.tree
```

and carry out midpoint rooting

```
phylip retree
```

before drawing the primate phylogeny

```
new2view outtree
```

to get

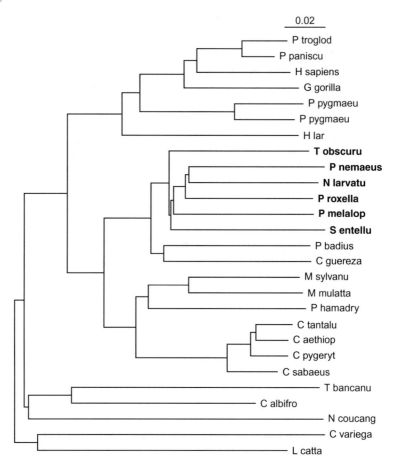

This is quite similar to the primate phylogeny recovered with UPGMA, though there are differences in the branching order. Most of these are in the clade shown in bold, where branch lengths are short. If branches are short, clustering is based on little information and hence subject to reinterpretation by different algorithms.

Answer 409 The number of ancestors, a, as a function of the number of generations, g, in a bi-parental genealogy is

$$a = 2^g.$$

Therefore, the (theoretical) number of ancestors 30 generations back is

$$2^{30} = \left(2^{10}\right)^3 \approx \left(10^3\right)^3 = 10^9.$$

This is much larger than the human population at that time: If a generation is 25 years, 30 generations take us back to 1267. The world population at that time was approximately 400 million; it took until the early nineteenth century for the human population to exceed 10^9.

Answer 410 We expect individual i_4 to have eight great grandparents in generation b_3, which is impossible with population size 7. So in small populations there is more sharing of ancestors than in large populations.

Answer 411 Blue individuals have left no descendants in the present. Red individuals are ancestors of all extant individuals. In contrast to these universal ancestors, black individuals have left some descendants in the present. Notice that as you go back in time, partial ancestors go extinct leaving only universal ancestors and non-ancestors [41]. From the point of view of the present, this means that eventually we all have the same ancestors. So next time someone tells you she is a descendant of X, who lived a long time ago, you know there are only two possibilities: Either this is true, then it is true for everyone, or, alas, it is false.

Answer 412 Set up the directory

```
mkdir Descent
cd Descent
```

Run the simulation using

```
  drawGenealogy -D 0 -i 4 -p 7 -C -g 7 -t testFig.tex
```

Typeset the figure and view it

```
latex testFig
dvips testFig -o -q
gv testFig.ps &
```

Answer 413 Simulate the times to the most recent universal ancestor

```
bash firstUnivAnc.sh > firstUnivAnc.dat
```

where `firstUnivAnc.sh` is

```
for a in 10 20 50 100 200 500 1000
do
    echo -n $a ' '
    for b in $(seq 100);
    do
        drawGenealogy -g 50 -c -p $a
    done |
        grep comm |
        awk '{s+=$8;c++}END{print s/c}'
done
```

Plot the simulation results

```
gnuplot -p plot1.gp
```

where `plot1.gp` is

```
set xlabel "Population Size"
set ylabel "Generations"
set pointsize 2
set log x
set key top center
f(x) = log(x) / log(2)
plot [][] "firstUnivAnc.dat" title "observed" w linesp,\
f(x) title "expected" wi li
```

The resulting graph shows the number of generations until the appearance of the first universal ancestor as a function of population size.

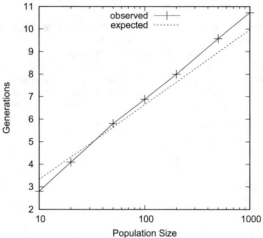

The expectation fits the simulation quite well.

Answer 414 Compute times until all present-day individuals have identical ancestors:

```
bash allUnivAnc.sh > allUnivAnc.dat
```

where `allUnivAnc.sh` is

```
for a in 10 20 50 100 200 500 1000
do
    echo -n $a ' '
    for b in $(seq 100);
    do
        drawGenealogy -g 50 -c -p $a
    done |
        grep iden |
        awk '{s+=$5;c++}END{print s/c}'
done
```

Plot the simulation together with the expectation

```
gnuplot -p plot2.gp
```

where `plot2.gp` is

```
set xlabel "Population Size"
set ylabel "Generations"
set pointsize 2
set log x
f(x) = 1.77 * log(x) / log(2)
plot [][] "allUnivAnc.dat" title "observed" wi linespoints,\
f(x) title "expected" wi li
```

to get the number of generations until all present-day individuals have identical ancestors as a function of population size.

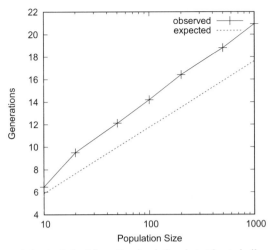

In this case, the expectation is distinct from the simulation, but at least similar.

Answer 415 The common ancestor of the two genes in i_4 lies beyond b_6. So in this uni-parental genealogy of genes, it takes much longer to reach the first common ancestors of genes than in the bi-parental genealogy of individuals.

Answer 416 Our extension of the Wright–Fisher model in Figure 6.4 looked like this; yours is bound to look different:

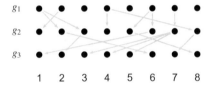

Answer 417 Yes, the Wright–Fisher simulation contains common ancestors, the most recent of which is marked in red:

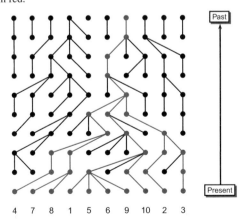

Answer 418 The commands

```
drawWrightFisher -p 10 -t wrapWf.tex -a -1
latex wrapWf
dvips wrapWf -o -q
```

generate Wright–Fisher simulations like

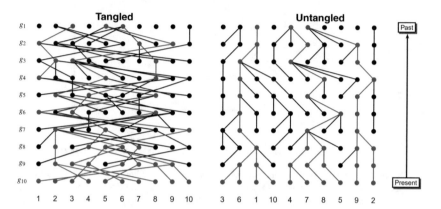

Among ten iterations, we found four instances with a time to the most recent common ancestor of ten generations or less.

Answer 419 The probability of two genes picking a common ancestor is $1/N \times 1/N = 1/N^2$. Since there are N opportunities for picking the same ancestor (each gene has one ancestor), the probability of any two genes picking the same ancestor is $1/N$.

Answer 420 In ten iterations of

```
awk -f trace1.awk -v seed=$RANDOM -v N=10       |
sort                                            |
uniq                                            |
wc -l
```

we found no instance where all ten lineages remained.

Answer 421 Compute P_n for $N = 10$:

```
BEGIN{
    n = 10
    pn = 1
    for(i=1; i<n; i++)
        pn *= 1 - i / n
    print pn
}
```

which returns $Pn \approx 4 \times 10^{-4}$. So we carried out 10^4 iterations of the simulation in Problem 420

```
for a in $(seq 10000)
do
    awk -f trace1.awk -v seed=$RANDOM -v N=10   |
        sort                                    |
        uniq                                    |
        wc -l
done                                            |
    awk 'BEGIN{c=0}{if($1==10)c++}END{print c}'
```

and found $P_n = 2 \times 10^{-4}$, which is close to the expected value.

Answer 422 We ran

```
awk -f trace2.awk -v seed=$RANDOM -v N=100 -v n=10
```

ten times and found three occasions where the number of lineages was reduced, that is, genes had picked common ancestors.

Answer 423 $P_a = 90/2000 = 0.045$. The simulation might look like this

```
for a in $(seq 1000)
do
    awk -f trace2.awk -v seed=$RANDOM -v N=1000 -v n=10
done |
    awk '{if($1 < 10)s++;c++}END{print s/c}'
```

which gave us $P_a = 0.048$, quite close to expected probability of an ancestor event.

Answer 424 Generate the number of lineages per generation

```
awk -f trace3.awk -v N=100 -v n=100 > trace3.dat
```

Plot the results

```
gnuplot -p plot3.gp
```

where plot3.gp is

```
set xlabel "Generations"
set ylabel "Lineages"
plot [][] "trace3.dat" t "" w l
```

to get

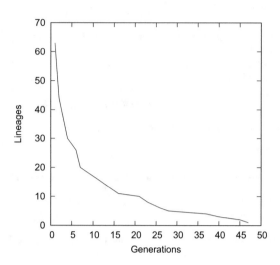

Answer 425 Commands like

```
awk -f trace3.awk -v seed=$RANDOM -v N=100 -v n=100
```

give widely varying times to the most recent common ancestor. Still, one might expect that
it takes much longer for 100 lineages to find their common ancestor than for two. However,
for individual runs of the simulation, times obtained with $n = 2$ are often quite similar as
those with $n = 100$.

Answer 426 $n = 2$: N generations; $n \rightarrow \infty$: $2N$ generations, that is to say, the expected
time to the most recent common ancestor, $E\{T_{\text{MRCA}}\}$, varies by no more than a factor of
2.

Answer 427 Run

```
bash simTmrca.sh > simTmrca.dat
```

where simTmrca.sh is

```
for i in 2 5 10 20
do
    echo -n ${i} ' '
    for j in $(seq 100)
    do
        awk -f trace3.awk -v seed=$RANDOM -v N=100 -v n=$i |
            tail -n 1
    done |
        awk '{s+=$1;c++}END{print s/c}'
done
```

and plot the result

```
plot -p plot4.gp
```

where plot4.gp is

```
set xlabel "Sample Size, n"
set ylabel "Average(T_{MRCA})"
f(x) = 2*100*(1-1/x)
set pointsize 2
set key top left
plot [][] "simTmrca.dat" t "Simulation" w linespoints,\
f(x) t "Expectation" w l
```

to get

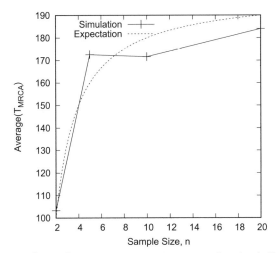

The simulated average time to the most recent common ancestor is quite similar to its expectation.

Answer 428 On the left is a example coalescent with all nodes labeled. It has four leaves, so it describes the genealogy of a sample of $n = 4$ genes:

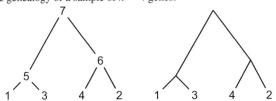

By convention, only leaves are labeled in a coalescent, as shown on the right.

Answer 429 T_1 would begin with the most recent common ancestor, the root of the coalescent, and go on forever. It is not shown, because the coalescent is bounded by the most recent common ancestor.

Answer 430 By running this code with i=4, i=3, and i=2, we got $T_4 = 0.10$, $T_3 = 0.03$, $T_2 = 0.29$; so the coalescence times were as follows:

Index	1	2	3	4	5	6	7
Node	1	2	3	4	5	6	7
Time	0.00	0.00	0.00	0.00	0.10	0.13	0.42

Time to the most recent common ancestor is the time of the root node, $0.42 \times 2N$ generations.

Answer 431 Code like

```
for a in $(seq 100)
do
    awk -v seed=$RANDOM -f genCoalTimes.awk -v n=1000 |
        tail -n 1
done |
    awk '{s+=$2;c++}END{print s/c}'
```

gave average times to the most recent common ancestor of $T_{\text{MRCA}} = 0.90$ for $n = 2$ and $T_{\text{MRCA}} = 1.98$ for $n = 1000$. Recall from Problem 426 that when measured in $2N$ generations, we expect

$$T_{\text{MRCA}} = \left(1 - \frac{1}{n}\right).$$

Answer 432 Here is a step-by-step depiction of shuffling, or put more formally, sampling without replacement; the values to be swapped are shown in bold. A particular position in the array might be picked repeatedly.

$r = 1, n = 5$	$r = 3, n = 4$	$r = 1, n = 3$	$r = 2, n = 2$	Result
1 2 3 4 5	1 2 3 4 5	1 2 3 4 5	1 2 3 4 5	1 2 3 4 5
1 2 3 4 5	5 2 **3** 4 **1**	**5** 2 **4** 3 1	4 **2** 5 3 **1**	4 2 5 3 1

Answer 433 When continuing the construction of the coalescent, for parent 6 we got as first child indexes 1 and 1. So the first child of node 6 is the node at position 1, which is 4. Then, node 3 is placed at position 1, and is thus drawn as the second child of 6. Finally, the node at position 1 is replaced by node 6:

Index	1 2 3 4 5 6 7
Node	1 2 3 4 5 6 7
	4 5
	3
	6
Child1	1 4
Child2	2 3
Time 0	0 0 0

This leaves only two children for 7, 5, and 6; so the final topology is as follows:

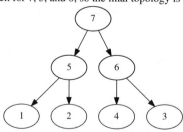

Answer 434 Solutions are bound to differ; ours is

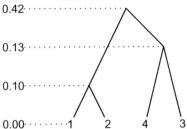

Answer 435 For a coalescent of sample size 4, the children of three internal nodes need to be picked:

```
awk -v seed=$RANDOM -v n=4 -f pickChildren.awk
# Pa C1    C2
5    1     2
6    1     2
7    1     1
```

So we can fill in our table:

Index	1	2	3	4	5	6	7
Node	1	2	3	4	5	6	7
	4	5					
	3	6					
	6						
Child1					1	4	3
Child2					2	5	6
Time	0.00	0.00	0.00	0.00	0.16	0.20	0.23

This gives the tree

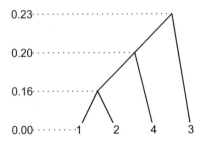

Answer 436 We label the branches on our coalescent and then draw the mutations:

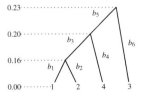

Branch	Length	Mutations
b_1	0.16	0
b_2	0.16	1
b_3	0.04	0
b_4	0.20	0
b_5	0.03	0
b_6	0.23	3

Our tree has a total of four mutations. This is far fewer than the corresponding expectation value:

```
watterson -n 4 -t 10
S = 18.333333
```

Answer 437 Here is an example run to generate two samples:

```
awk -f coalescent.awk -v seed=$RANDOM -v theta=10 -v
      sampleSize=4 -v numSamples=2
S= 28
S= 9
```

S is the number of mutations (segregating sites) found in a simulated sample.

Answer 438 The command

```
awk -f coalescent.awk -v seed=$RANDOM -v theta=10 -v
      sampleSize=4 -v numSamples=1 -v printTree=1 |
grep T              |
new2view
```

returns, for example

(a)   **(b)**   **(c)**

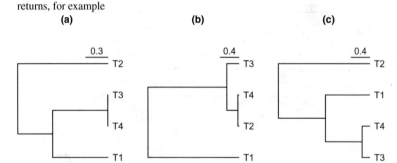

The coalescent time is measured in units of $2N$ generations. Accordingly, in coalescent **(a)** the scale refers to $0.3 \times 2N$ generations in a haploid population.

Answer 439 Our coalescent simulation looks like this:

```
ms 4 10000 -t 10     |
grep '^s'            |
awk '{s+=$2;c++}END{print s/c}'

18.32
```

The expectation was already computed in Problem 436, 18.33; this is very close to the simulated value.

Answer 440

```
ms 4 10000 -t 10     |
grep '^s'            |
awk '{if($2<=1)s++;c++}END{print "P="s/c}'
```

```
P=0.0102
```

Our test of the Wright–Fisher model rejects the null hypothesis with an error probability of just 1%. So the difference between observation and expectation is significant if we apply the common threshold of $\alpha = 0.05$.

Answer 441 First position:

```
tabix http://guanine.evolbio.mpg.de/problemsBook/chr19.mgp.
    vcf.gz 19:1-5000000 |
head -n 1
```

```
19   3078554
```

Last position:

```
tabix http://guanine.evolbio.mpg.de/problemsBook/chr19.mgp.
    vcf.gz 19:50000000-70000000 |
tail -n 1
```

```
19   61331223
```

Answer 442

```
tabix http://guanine.evolbio.mpg.de/problemsBook/chr19.mgp.
    vcf.gz 19:3078554-61331223 |
wc -l
```

```
1617408
```

That is, there are

$$\frac{1,617,408}{61,331,223 - 30,78,554 + 1} \approx 0.0278$$

SNPs per position. 17 mice have 34 chromosomes; we can compute the corresponding harmonic number $\sum_{i=1}^{33} 1/i$ using

```
watterson -t 1 -n 34
```

```
S = 4.088798
```

So the per nucleotide population mutation rate

$$\theta = \frac{0.0278}{4.0888} = 0.0068.$$

Answer 443 Count SNPs on mouse chromosome 19:

```
tabix http://guanine.evolbio.mpg.de/problemsBook/chr19.mgp.
    vcf.gz 19:5,189,001-5,190,000 |
wc -l
```

```
40
```

We get the θ for 1 kb from the per nucleotide value of 0.0068 and compute the expected number of SNPs:

```
watterson -t 6.8 -n 34
```

```
S = 27.803828
```

Significance

```
ms 34 10000 -t 6.8   |
grep '^s'            |
awk '{c++;if($2>=40)s++}END{print "P="s/c}'
```

```
P=0.1227
```

So the difference between theory and observation is not significant and we do not reject the null hypothesis, the Wright–Fisher model.

Answer 444 We search for *Plin5* in both data sets:

```
grep Plin5 all_a.txt > plin5_a.txt
grep Plin5 all_b.txt > plin5_b.txt
```

Answer 445 This can be solved in various ways, including mental arithmetic, here is our solution:

```
awk '{for(i=2;i<=NF;i++){s+=$i;c++}}END{print s/c}' plin5_a.txt
11.9303
```

and

```
awk '{for(i=2;i<=NF;i++){s+=$i;c++}}END{print s/c}' plin5_b.txt
12.947
```

Answer 446 We run the program

```
testMeans plin5_a.txt plin5_b.txt
Plin5 1.193e+01 1.295e+01 7.871e-03
```

The first two numbers are the averages we already computed, the third number is the P-value, that is, the error probability when rejecting the null hypothesis that the difference between the two samples is negligible. If we use the customary cutoff value of $\alpha = 0.05$, the difference is significant.

Answer 447 We aimed for very high precision and ran the Monte Carlo test with 10^6 iterations:

```
testMeans -t m -i 1000000 plin5_a.txt plin5_b.txt
Plin5 1.193e+01 1.295e+01 7.934e-03
```

This still varies between different runs, but it is quite close to the 7.871×10^{-3} computed previously.

Answer 448 We simulate the data

```
simNorm -m 12 -i 100 > experiment1.txt
simNorm -m 12 -i 100 > experiment2.txt
```

Then we compute the P values and count the cases where $P \leq 0.05$:

```
testMeans experiment1.txt experiment2.txt |
awk '{if($4<=0.05)print}'                 |
wc -l
6
```

In this case, the observed false-positive rate is $6/100 = 0.06$. This is close to the expected $\alpha = 0.05$, but is bound to vary between runs.

Answer 449 We ran

```
simNorm -m 12 -i 10000 > experiment1.txt
simNorm -m 12 -i 10000 > experiment2.txt
testMeans experiment1.txt experiment2.txt |
awk '{if($4<=0.05)print}' | wc -l
462
```

So the observed false-positive rate is $462/10000 = 0.0462$, which again is close to the expected 0.05.

Answer 450 We compute

$$1 - (1 - 0.05)^{100} = 0.994;$$

that is to say, there is a 99.4% chance of getting at least one false-positive when carrying out 100 hypothesis tests with $\alpha = 0.05$.

Answer 451 We use the same computation of the false-positive rate as before, except that this time we divide α by 10^4:

```
testMeans experiment1.txt experiment2.txt |
awk '{if($4<=0.05/10000)print}'           |
wc -l
0
```

In other words, after Bonferroni correction, we found not a single false-positive result, the type I error was all but eliminated (again, your result may differ slightly). We know that all samples were drawn from the same population, so this is the correct result.

Answer 452 Simulate the data

```
simNorm -m 6 -d 2.5 -i 10000 > experiment1.txt
simNorm -m 8 -d 2.5 -i 10000 > experiment2.txt
```

and determine the frequency with which the null hypothesis is *not* rejected:

```
testMeans experiment1.txt experiment2.txt |
awk '{if($4>0.05)print}'                  |
wc -l
6866
```

Since the samples were drawn from populations with different μ, every acceptance of the null hypothesis is a false-negative result: $\beta = 6866/10000 \approx 0.68$. In other words, the false-negative rate is large if the difference in means, also called the "effect size", is small.

Answer 453 Simulate the data

```
simNorm -m 6 -d 2.5 -i 10000 > experiment1.txt
simNorm -m 12 -d 2.5 -i 10000 > experiment2.txt
```

and again determine the frequency with which the null hypothesis is *not* rejected:

```
testMeans experiment1.txt experiment2.txt |
awk '{if($4 > 0.05)s++;c++}END{print "beta=" s/c}'
beta=0.0074
```

This means the false-negative rate drops dramatically when the effect size is increased.

Answer 454 Simulate the data

```
simNorm -m 6 -d 2.5 -i 10000 > experiment1.txt
simNorm -m 12 -d 3.5 -i 10000 > experiment2.txt
```

and compute the false-negative rate:

```
testMeans experiment1.txt experiment2.txt |
awk '{if($4 > 0.05)s++;c++}END{print "beta=" s/c}'
beta=0.0489
```

Answer 455 Apply the corrected α:

```
testMeans experiment1.txt experiment2.txt |
awk '{if($4 > 0.05/10000)s++;c++}END{print "beta=" s/c}'
beta=0.9811
```

This means, the Bonferroni correction leads to a large type II error rate and thereby obscures almost entirely the true difference between the two sets of experiments.

Answer 456 As before we count the nonsignificant P values:

```
testMeans experiment1.txt experiment2.txt |
sort -g -k 4                                |
awk '{if($4>c*0.05/10000)s++; c++}END{print "beta=" s/c}'
beta=0.0536
```

Here, the β is very close to $\delta = 0.05$, as it should be.

Answer 457 Simulate the data

```
simNorm -i 10000 -m 6 -d 2.5 > experiment1.txt
simNorm -i 10000 -m 6 -d 2.5 > experiment2.txt
```

and analyze them as before

```
testMeans experiment1.txt experiment2.txt       |
cut -f 4                                         |
sort -g                                          |
awk 'BEGIN{m=10000;d=0.05}{if($1<=NR*d/m)print $1}' |
wc -l
0
```

The type I error seems to have been removed as thoroughly as with the Bonferroni correction, but without the concomitant increase in type II error.

Answer 458 To test the effect of sample size, you could execute

```
bash sim.sh |
gnuplot -p plot.gp
```

where `sim.sh` is

```
for a in 2 5 10 20 50 100 200 500
do
    echo -n $a ' '
    simNorm -i 10000 -m 6 -d 2.5 -n $a > experiment1.txt
    simNorm -i 10000 -m 7 -d 2.5 -n $a > experiment2.txt
    testMeans experiment1.txt experiment2.txt          |
    cut -f 4                                            |
    sort -g                                             |
    awk '{if($1<=NR*0.05/10000)c++}END{print 1-c/10000}'
done
```

and `plot.gp`

```
set xlabel "Sample Size (n)"
set ylabel "Type II Error (beta)"
plot "< cat" t "" w l
```

to get

The type II error decreases dramatically with increasing sample size. We say the power of the test grows. However, increasing the sample size beyond 20 does not improve its power substantially any more.

Answer 459 To obtain the type II error as a function of sample size, execute

```
bash sim2.sh |
gnuplot -p plot2.gp
```

where `sim2.sh` is

```
for a in 2 5 10 20 50 100 200 500
do
    echo -n $a ' '
    simNorm -i 10000 -m 6 -d 2.5 -n $a > experiment1.txt
    simNorm -i 10000 -m 7 -d 2.5 -n $a > experiment2.txt
    testMeans experiment1.txt experiment2.txt          |
    cut -f 4                                            |
    sort -g                                            |
    awk '{if($1<=NR*0.05/10000)c++}END{print 1-c/10000}'
done
```

and `plot2.gp`

```
set xlabel "Sample Size (n)"
set ylabel "Type II Error (beta)"
f(x)=0.05
plot "< cat" t "" w l, f(x) t "" w l
```

to get

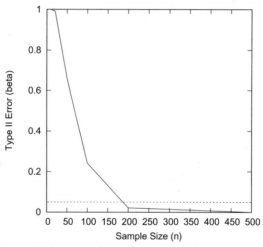

A sample size of $n \approx 200$ is now needed to obtain $\beta < 0.05$.

Answer 460 The number of experiments is the number of rows in `all_a.txt` or `all_b.txt`:

```
wc -l all_a.txt
25789 all_a.txt
```

The number of distinct genes is

```
cut -f 1 all_a.txt | sort | uniq | wc -l
14779
```

Answer 461

```
testMeans all_a.txt all_b.txt                        |
sort -k 4 -g                                         |
awk 'BEGIN{m=25789;d=0.1}{if($4<=NR*d/m)print}'      |
cut -f 1                                             |
sort                                                 |
uniq > genes.txt
```

The file `genes.txt` contains 209 distinct gene identifiers.

Answer 462 The highest red node is "Metabolic Process". The underlying study is concerned with the effect of fatty food on mice. Apparently, a change in food leads to a change in metabolism, which makes sense.

Answer 463 Do not forget to construct the new directory `RelationalDb` to keep your work on relational databases separate from the rest. The file `fatty_food.sql` contains

```
create table fatty_food(
   Sym varchar(18),
   M1 float,
   M2 float,
   M3 float,
   M4 float,
   M5 float,
   M6 float,
   M7 float,
   M8 float,
   primary key(Sym)
);
```

Notice that capital and lower case letters are not distinguished in SQL.

Answer 464 The file `fatty_food.sql` now contains

```
create table fatty_food(
   Sym varchar(18),
   M1 float,
   M2 float,
   M3 float,
   M4 float,
   M5 float,
   M6 float,
   M7 float,
   M8 float,
   primary key(Sym),
   foreign key (Sym) references normal_food(Sym)
);
```

Answer 465 The following commands construct the database

```
sqlite3 mouseExpress.db
.read normal_food.sql
.read fatty_food.sql
.separator "\t"
```

```
.import normal_food.txt normal_food
.import fatty_food.txt fatty_food
```

Notice that `normal_food` needs to be filled before `fatty_food`; otherwise, the foreign key constraint is violated.

Answer 466 Here is the required combination of `insert`, `select`, and `delete` commands:

```
insert into normal_food
values('toy_gene2',17.1, 9.5, 27.7, 6.5, 24.1, 30.2,
30.6, 14.3);
select * from normal_food where sym like 'toy_gene2';
delete from normal_food where sym like 'toy_gene2';
select * from normal_food where sym like 'toy_gene2';
```

Answer 467 If we enter

```
insert into normal_food
values('Plin5',17.1, 9.5, 27.7, 6.5, 24.1, 30.2,
30.6, 14.3);
insert into fatty_food
values('Plin5',17.1, 9.5, 27.7, 6.5, 24.1, 30.2,
30.6, 14.3);
```

we get the error messages

```
Error: near line 1: UNIQUE constraint failed: normal_food.sym
Error: near line 4: UNIQUE constraint failed: fatty_food.sym
```

Answer 468 We type

```
insert into fatty_food
values ('toy_gene3',3.4, 8.0, 4.4, 26.7, 8.6, 26.6, 4.8, 20.5);
```

and get the error message

```
FOREIGN KEY constraint failed
```

as there is no entry for `toy_gene3` in `normal_food`. If you do not get this error, you probably did not enter

```
PRAGMA foreign_keys = ON;
```

in this database session.

Answer 469 We write

```
.read insert.sql
```

where `insert.sql` contains the same commands we entered interactively, and so we get the same result.
To enter one value too may or too few, we can use

```
-- Insert too few values
insert into normal_food
values('toy_gene3',17.1, 9.5, 27.7, 6.5, 24.1, 30.2, 30.6);
-- Insert too many values
insert into normal_food
values('toy_gene3',17.1, 9.5, 27.7, 6.5, 24.1, 30.2, 30.6,
    14.3, 16.7);
```

to trigger the error messages

```
Error: near line 2: table normal_food has 9 columns but 8
    values were supplied
Error: near line 6: table normal_food has 9 columns but 10
    values were supplied
```

Notice that comments are marked by -- in SQL.

Answer 470 Counting in SQL:

```
select count(*) from normal_food;
14779
select count(*) from fatty_food;
14779
```

Answer 471

```
select sym, (m1+m2+m3+m4+m5+m6+m7+m8)/8 from normal_food
    limit 3;
Nppa    6.37625
Gm12689 6.791625
Hvcn1   6.951
```

Answer 472 Compute the maximum

```
select sym, max((m1+m2+m3+m4+m5+m6+m7+m8)/8) from normal_food;
Pigt    19.068
```

the minimum

```
select sym, min((m1+m2+m3+m4+m5+m6+m7+m8)/8) from normal_food;
Trpv3   5.817125
```

and, finally, the average

```
select avg((m1+m2+m3+m4+m5+m6+m7+m8)/8) from normal_food;
8.49976156201365
```

Answer 473 Compute the maximum

```
select sym, max((m1+m2+m3+m4+m5+m6+m7+m8)/8) from fatty_food;
Pigt    19.068
```

the minimum

```
select sym, min((m1+m2+m3+m4+m5+m6+m7+m8)/8) from fatty_food;
Trpv3   5.823
```

and the average

```
select avg((m1+m2+m3+m4+m5+m6+m7+m8)/8) from fatty_food;
8.50960544522628
```

Answer 474

```
select normal_food.sym,
        (normal_food.m1 + normal_food.m2 +
        normal_food.m3 + normal_food.m4 +
        normal_food.m5 + normal_food.m6 +
        normal_food.m7 + normal_food.m8) / 8,
        (fatty_food.m1 + fatty_food.m2 +
        fatty_food.m3 + fatty_food.m4 +
        fatty_food.m5 + fatty_food.m6 +
        fatty_food.m7 + fatty_food.m8) / 8
from normal_food join fatty_food using (sym);
```

During construction of this command, you might have restricted its output by adding

```
limit 10
```

at the end. However, for the next Problem we need the full output, so make sure `join.sql` is saved as shown above.

Answer 475

```
sqlite3 mouseExpress.db < join.sql |
tr '|' '\t'                         |
awk -f fc.awk
Hsd3b5 1.3583
```

Write the AWK program `fc.awk` to find the gene with the largest fold change. where `fc.awk` contains

```
BEGIN{
    max = -1
}
{
    if($2 > $3 && $3 > 0)
        fc = $2 / $3
    else if($2 > 0)
        fc = $3 / $2
    else
        fc = -1
    if(fc > max){
        max = fc
        sym = $1
    }
}
END {
    print sym, max
}
```

Answer 476

```
sqlite3 mouseExpress.db < join.sql | tr '|' '\t' > avg.txt
```

Answer 477 Here is our solution:

```java
import java.sql.*;
public class MouseExpressDb2{
    public static void main( String args[] ){
        String query = "select * from normal_food join
            fatty_food using(sym)";
        double fc, a1, a2, max;
        String name = null;
        try{
            Class.forName("org.sqlite.JDBC");
            Connection c = DriverManager.getConnection("jdbc:
                sqlite:mouseExpress.db");
            Statement stmt = c.createStatement();
            ResultSet rs = stmt.executeQuery(query);
            max = -1;
            while(rs.next()){
                a1 = a2 = 0.0;
                for(int i=2;i<=9;i++){
                    a1 += rs.getFloat(i);
                    a2 += rs.getFloat(i+8);
                }
                a1 /= 8;
                a2 /= 8;
                if(a1 > a2 && a2 > 0)
                    fc = a1 / a2;
                else if(a1 > 0)
                    fc = a2 / a1;
                else
                    fc = -1;
                if(fc > max){
                    max = fc;
                    name = rs.getString(1); // Access column #1,
                            i.e.sym, a string
                }
            }
            System.out.printf("%s\t%.3f\n", name, max);
        }catch(Exception e){
            System.err.println(e.getClass().getName() + ": "
                + e.getMessage());
            System.exit(0);
        }
    }
}
```

When we run this, we get the same result as with `fc.awk`

```
java -cp sqlite-jdbc-3.15.1.jar:. MouseExpressDb2
Hsd3b5  2.675
```

Answer 478 At the time of writing ENSEMBL consisted of

```
mysql -h ensembldb.ensembl.org -u anonymous -e "show databases" |
tail -n +2                                                       |
wc -l
6346
```

Answer 479 At the time of writing the latest mouse `core` database in ENSEMBL was

```
mysql -h ensembldb.ensembl.org -u anonymous -e "show databases" |
grep mus_musculus_core                                           |
tail -n 1
mus_musculus_core_88_38
```

Answer 480

```
mysql -h ensembldb.ensembl.org -u anonymous -D
    mus_musculus_core_88_38 -e "show tables" |
tail -n +2 |
wc -l
73
```

Answer 481 Enter

```
mysql ... -e "describe seq_region"
```

To get

```
+-----------------+------------------+------+-----+---------+----------------+
| Field           | Type             | Null | Key | Default | Extra          |
+-----------------+------------------+------+-----+---------+----------------+
| seq_region_id   | int(10) unsigned | NO   | PRI | NULL    | auto_increment |
| name            | varchar(255)     | NO   | MUL | NULL    |                |
| coord_system_id | int(10) unsigned | NO   | MUL | NULL    |                |
| length          | int(10) unsigned | NO   |     | NULL    |                |
+-----------------+------------------+------+-----+---------+----------------+
```

In `mysql`, "attributes" are called "Field", and hence the first column of this table is the one most relevant for us.

Answer 482 We pipe the command given in the problem through

```
awk '{s+=$2}END{print s}'
```

to get a genome length of 2,725,521,370 bp.

Answer 483 The command

```
mysql ... -e "describe exon"
```

gives

```
+-------------------+----------------------+------+-----+---------+----------------+
| Field             | Type                 | Null | Key | Default | Extra          |
+-------------------+----------------------+------+-----+---------+----------------+
| exon_id           | int(10) unsigned     | NO   | PRI | NULL    | auto_increment |
| seq_region_id     | int(10) unsigned     | NO   | MUL | NULL    |                |
| seq_region_start  | int(10) unsigned     | NO   |     | NULL    |                |
| seq_region_end    | int(10) unsigned     | NO   |     | NULL    |                |
| seq_region_strand | tinyint(2)           | NO   |     | NULL    |                |
| phase             | tinyint(2)           | NO   |     | NULL    |                |
| end_phase         | tinyint(2)           | NO   |     | NULL    |                |
| is_current        | tinyint(1)           | NO   |     | 1       |                |
| is_constitutive   | tinyint(1)           | NO   |     | 0       |                |
| stable_id         | varchar(128)         | YES  | MUL | NULL    |                |
| version           | smallint(5) unsigned | YES  |     | NULL    |                |
| created_date      | datetime             | YES  |     | NULL    |                |
| modified_date     | datetime             | YES  |     | NULL    |                |
+-------------------+----------------------+------+-----+---------+----------------+
```

where again the first column contains the information we are looking for.

Answer 484 Compute the number of nucleotides contained in exons:

```
mysql ... -e "select sum(seq_region_end - seq_region_start + 1)
   from exon"
+------------------------------------------+
| sum(seq_region_end - seq_region_start + 1) |
+------------------------------------------+
|                               168950237  |
+------------------------------------------+
```

Then, divide this number by the total genome length

$$\frac{168,950,237}{2,725,521,370} \times 100 \approx 6.2\%$$

6.2 % of the mouse genome is covered by exons.

Answer 485 For this, we need the attribute `is_constitutive`:

```
mysql ... -e "select sum(seq_region_end - seq_region_start + 1)
   from exon where is_constitutive=1"

+------------------------------------------+
| sum(seq_region_end - seq_region_start + 1) |
+------------------------------------------+
|                               38483186  |
+------------------------------------------+
```

Hence, $38,483,186/2,725,521,370 \times 100 \approx 1.4\%$ of the mouse genome is covered by constitutive exons.

Answer 486 We use

```
select display_label, description
from xref
where display_label like 'Hsd3b5'
```

to learn that *Hsd3b5* is a hydroxy-delta-5-steroid dehydrogenase, 3 beta- and steroid delta-isomerase 5.

Answer 487 To find the `stable_id` of *Hsd3b5*, we can use

```
select display_label, gene.stable_id
from xref join gene on display_xref_id = xref_id
where display_label like  'Hsd3b5'
```

and get the gene ENSMUSG00000038092. All mouse genes have a `stable_id` of the form ENSMUSG...

Answer 488 Join the `gene` and `transcript` tables via `gene_id` and filter for ENS-MUSG00000038092:

```
select gene.stable_id, transcript.stable_id
from gene join transcript using(gene_id) where
     gene.stable_id like 'ENSMUSG00000038092'
```

which gives ENSMUST00000044094, that is one transcript. All mouse transcripts have a `stable_id` of the form ENSMUST... You might have been tempted to join the two tables via `stable_id`, which is another attribute common to `gene` and `transcript`. However, the `stable_ids` in gene have the format ENSMUSG..., while we already saw that the `stable_ids` of transcripts are ENSMUST... Connecting the two tables is thus only possible via `gene_id`.

8.2 Appendix: UNIX Guide

This is a brief summary of the most essential UNIX commands. It is meant to be read next to a computer running the UNIX operating system, so that readers can experiment. For further reading, we recommend the system's online documentation and [1].

8.2.1 File Editing

Of the many text editors available for UNIX systems, we recommend `emacs`, as it comes with a standard graphical user interface for casual use. At the same time, it is a powerful and versatile tool used by many IT professionals. As a result, it is available on many systems. It is started by typing

```
emacs &
```

This opens a window with standard menus running emacs. A new file is opened by clicking on the corresponding icon. It is then edited using standard keyboard input and mouse moves. In addition, emacs comes with a rich set of keyboard shortcuts, called "key bindings", making its use much more efficient than with these traditional techniques. Table 8.1 lists the key bindings we use regularly in our own work. The table also illustrates the principle of creating different versions of commands through alternate use of the control key and the meta key. Commands referring to sentences only work if a full stop is followed by at least two blanks. Two of these key combinations are special: C-x and M-x. C-x is a prefix for other key combinations, and we have listed the ones we find most useful in Table 8.2. M-x is also a prefix for further commands, but these are called by extended names like calendar rather than one or two characters. Again, Table 8.2 lists our favorites. The full list of key bindings can be accessed by C-h b. The best place to start using the key bindings is the "Emacs Tutorial" opened by C-h t.

8.2.2   Working with Files

The many things we need to do with text files on a regular basis include viewing them, measuring their size, finding patterns in them, and so on. Table 8.3 lists some of the commands to carry out these routine tasks. It is a good idea to learn at least some of them by heart, as they are used constantly when working with the UNIX command line.

UNIX commands tend to come with numerous options. These are documented in the manual pages, which are accessed using the program man; for example,

```
man ls
```

This invokes a text viewer responsive to some of the same key bindings for cursor movement as emacs, e.g., C-v to scroll down and M-v to scroll up. Inside man, h invokes help and q quits.

8.2.3   Entering Commands Interactively

Any command entered at a command prompt is interpreted by a piece of software called the "shell", which runs inside a terminal window. There are different kinds of shells and the command

```
echo $SHELL
```

returns the active shell. The following description applies to the bash shell. If it is not already running, it can be started by entering

```
bash
```

Table 8.1 Paired emacs key bindings (shortcuts)

Key	Binding	Key	Binding
C-a	Move to beginning of line	M-a	Move to beginning of sentence
C-b	Move backward one character	M-b	Move backward one word
C-d	Delete character	M-d	Delete word to the right
—	—	M-BACKSP	Delete word to the left
C-e	Move to end of line	M-e	Move to end of sentence
C-f	Forward one character	M-f	Forward one word
C-g	Keyboard quit	—	—
C-h	Help	M-h	Mark paragraph
C-k	Delete line	M-k	Delete sentence
C-l	Center buffer on current line	M-l	Lower case word
C-n	Next line	—	—
C-p	Previous line	—	—
—	—	M-q	Layout paragraph
C-r	Search backward	M-r	Move to top/bottom of window
C-s	Search forward	—	—
C-t	Transpose characters	M-t	Transpose words
—	—	M-u	Upper case word
C-v	Scroll up	M-v	Scroll down
C-w	Delete selection	M-w	Copy selection
C-x	Command prefix (Table 8.2)	M-x	Execute extended command (Table 8.2)
C-y	Paste	—	—
C-z	Suspend frame	M-z	Delete to character
C-_	Undo	—	—
C-SPC	Set mark	—	—
C-+	Increase font size	—	—
C-	Decrease font size	—	—
—	—	M-<	Move to beginning of buffer
—	—	M->	Move to end of buffer

The command line completes prefixes of commands and file names in response to pressing TAB once if the prefix is unique. Otherwise, by pressing TAB a second time, the list of possible completions is presented. The most effective way of interacting with the command line is to let the autocompletion do as much work as possible by carefully mixing typing and tabbing. With a little practice this becomes second nature.

Like man, the shell is responsive to the same basic key bindings as emacs, which is an added benefit from learning them.

Table 8.2 A selection of frequently used composite commands in `emacs`

Key	Binding
`C-x b`	Switch buffer
`C-x k`	Kill buffer
`C-x o`	Switch to other buffer
`C-x C-c`	Quit
`C-x C-f`	Find file
`C-x C-s`	Save buffer
`C-x C-w`	Write file
`C-M-\`	Indent region
`M-x g`	Go to line
`M-x help`	Start help menu
`M-x shell`	Run shell in `emacs` buffer
`M-x count-words`	Count lines, words, and characters
`M-x rename-buffer`	Rename the current buffer
`M-x calendar`	Start calendar

8.2.4 Combining Commands: Pipes

UNIX commands such as those listed in Table 8.3 can be combined into programs by using the output from one command as the input for another. To do this, individual commands are combined via *pipes* denoted by a vertical line, "`|`". For example, to count the number of files,

```
ls | wc -l
```

where the output of `ls` is the input of `wc`.

The shell can also expand file names. Hence,

```
ls *.txt | wc -l
```

returns the number of files in the working directory that end in `.txt`.

8.2.5 Redirecting Output

By default, the result of a command is printed to the screen standard output stream called stdout. This usually corresponds to the screen. Output can be redirected to a file by writing, for example,

```
ls > tmp
```

which creates the file `tmp` (check using `ls`) and writes the results of the command `ls` into it (check using `cat tmp`). The simple redirection command, >, overwrites

Table 8.3 Commands for working with files

Command	Explanation
cat	Print (conCATenate) to screen
-n	Print numbered lines
-b	Print non-blank lines
-s	Squeeze blank lines
cp *file1* *file2*	Copy *file1* to *file2*, overwrite old *file2* if it exists
cp *file1* *file2* *toDir*	Copy *file1* and *file2* to directory *toDir*
cut -f n	Cut the nth field
diff *fromFile* *toFile*	Find differences between *fromFile* and *toFile*
grep *pattern*	Print lines matching *pattern*
-v	Print lines not matching
-A n	Print matching lines and n lines after
-B n	Print matching lines and n lines before
head *filename*	Print first 10 lines of file
-n n	Print first n lines
join *file1* *file2*	Join two sorted files on 1st column
less	Pager for viewing text
ls	List names of all files in current directory
-l	Long listing for more information
-r	List in reverse order
-t	List in time order, most recent first
-u	List by time last used
mv *file1* *file2*	Move *file1* to *file2*, overwrite old *file2* if it exists
rm *filenames*	Remove named files, irrevocably
-r	Remove recursively directories and their contents
rmdir *directory*	Remove named directory
sort	Sort files alphabetically by line
-n	Sort numerically
-k n	Sort by column n
-r	Reverse sort
-R	Randomize
tac	Reverse lines of named files
tail	Print last 10 lines of file
-n	Print last n lines
+n n	Start printing file at line n
uniq *filenames*	Filter out repeated lines in sorted input
-c	Count repeated lines
-d	Print duplicated lines
-u	Print unique lines
wc	Count lines, words, and characters
-l	Count lines
-L	Return length of the longest line

the original content of the target file. The variant >> appends to whatever is already in the file:

```
ls >> tmp
```

The redirection can also go the other way:

```
cat < tmp
```

This writes the contents of `tmp` to the standard input, from where they can be read by `cat`.

8.2.6 Shell Scripts

Any command entered on the command line can also be submitted to the system from a text file called a "shell script". Suppose the file `ls.sh` contains a single line,

```
ls
```

This can be executed by passing it to the `bash`:

```
bash ls.sh
```

Alternatively, `ls.sh` can be made executable

```
chmod +x ls.sh
```

and then run

```
./ls.sh
```

Shell scripts can contain do loops and conditional statements. Say, a set of sequence files with names of the form `fileName.txt` need changing to `fileName.fasta`. This script generates ten example files:

```
for f in 1 2 3 4 5 6 7 8 9 10
do
    echo '>s'${f} > s${f}.txt
    echo 'ACCGT' >> s${f}.txt
done
```

The variable `f` takes the values of the list to the right of `in`. The value of `f` is referenced by writing it in curly brackets and prefixing it with `$`. The command `echo` prints its argument. These files are in *fasta* format, a header line starting with `>` followed by one or more lines of sequence data.

Instead of explicitly specifying a sequence of numbers, `seq` can be used:

```
for f in $(seq 10)
```

To change the file extensions from `.txt` to `.fasta`, we construct the script `rename.sh`, which allows us to write

```
bash rename.sh *.txt
```

where `rename.sh` is

```
for f in $@
do
     mv ${f}   ${f%txt}fasta
done
```

Notice the `for`-construct:

```
for f in $@
```

which loops over every argument on the command line, in our case all files ending with `.txt`. Further, in the construct

```
${f%txt}fasta
```

the suffix `txt` of the file name is deleted with the `%` operator and then replaced by `fasta`.

8.2.7   Directories

Directories may contain files and other directories. The UNIX file system is thus hierarchical and can be depicted as a tree of directories. Figure 8.1 shows a portion of this tree. There are three special types of directories:

1. Root directory: The most basic directory situated at the root of the file system (Fig. 8.1). Its name is /
2. Home directory: The directory accessed after logging in. It is the only directory in which a user can create files and directories.
3. Working directory: The directory a user is currently working in.

   The full name of a directory such as

$$/home/tom$$

is known as its *path*. A forward slash (/) can therefore either refer to the root directory or function as a delimiter of directory names. A variation of the name of the root directory, ~/, refers to the user's home directory. Table 8.4 summarizes the essential commands for working with directories.

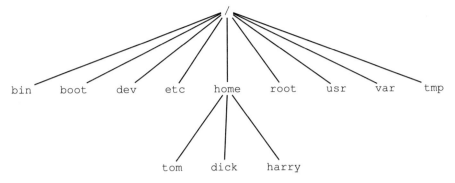

Fig. 8.1 A sample of the directories that make up a typical UNIX file system

Table 8.4 Commands for working with UNIX directories

Command	Explanation
pwd	Print working directory
cd	Return to home directory
cd *directoryname*	Change to the named directory
cd ..	Move up one level in the file system
mkdir *directoryname*	Make the named directory
rmdir *directoryname*	Remove the named directory

```
a 10
b 11
c 12
d 15
e 11
```

Fig. 8.2 Sample data contained in the file data.dat

8.2.8 *Filters*

Programs for filtering textual data are the bread and butter of practical bioinformatics. The following sections introduce four of the most popular filters: grep, tr, sed, and awk. All three make use of a common notation for specifying patterns in strings. Such patterns are called "regular expressions". We give examples of them when dealing with the individual filters and explain them in more detail afterward.

grep

Consider the data shown in Fig. 8.2, and say we wanted to extract all lines matching 11. The command

```
man ls
```

where data.dat contains the listing in Fig. 8.2 would return

```
b 11
e 11
```

With -v the lines *not* matching are printed:

```
grep -v 11 data.dat
a 10
c 12
d 15
```

Grep can also be used to search for more complex patterns specified as regular expressions. For example, [25] matches "2" or "5":

```
grep [25] data.dat
c 12
d 15
```

tr

The program tr is used to translate or delete characters. For example, the command

```
tr -d '\n' < data.dat
```

removes all line breaks. Notice that unlike many UNIX tools, tr does not take file names as arguments. Hence, the < notation for reading from data.dat. Alternatively, we could have written

```
cat data.dat | tr -d '\n'
```

To "translate" tabulators into newlines, we enter

```
tr '\t' '\n' < data.dat
```

Or to convert the line labels to upper case, we can write

```
tr a-z A-Z < data.dat
```

where a-z is a character range indicating any lower case letter. Number ranges are coded similarly, which would allow us to express numbers as characters:

```
tr 0-9 a-z < data.dat
```

A biologically more relevant translation would be to encode A or G as purine (R) and C or T as pyrimidine (Y):

```
echo ACGT | tr AG R | tr CT Y
```

sed

Emacs is an interactive editor. In contrast, sed is a noninteractive, "stream" editor. Perhaps, its simplest operation is to delete a single line, say the second line:

```
sed '2d' data.dat
```

Instead of deleting the second line, it can be printed

```
sed '2p' data.dat
```

By default sed applies its pattern to every line it encounters and prints all other lines unchanged. So in the previous example, the line

```
b 11
```

was printed twice. The option −n restricts the output to the matching lines:

```
sed -n '2p' data.dat
```

It is also possible to print line ranges:

```
sed -n '2,4p' data.dat
```

A replacement for head would be

```
sed -n '1,10p' data.dat
```

Like directories in paths, regular expressions are delineated by pairs of /

```
sed '/b/p' data.dat
```

So

```
sed -n '/b/p' data.dat
```

is an alternative to

```
grep b data.dat
```

The complement version of grep,

```
grep -v b data.dat
```

becomes in sed

```
sed '/b/d' data.dat
```

Apart from the print and delete operations, which in practice are often dealt with using grep, sed can carry out substitutions, which are beyond the capabilities of grep; for example,

```
sed 's/1/9/' data.dat
```

substitutes the first occurrence of a "1" by a "9" in each line. The *global* version of this command replaces all occurrences of "1" by "9",

```
sed 's/1/9/g' data.dat
```

The regular expression first encountered with `grep`, `[25]`, which specifies 2 *or* 5, can also be used in a substitution

```
sed 's/[25]/X/g' data.dat
```

AWK

AWK mixes a programming language with the line- and pattern-based paradigm of `grep` and `sed`. The following exposition is adapted from the original description of the language by its authors [4, Appendix A]. An AWK program is executed as

```
awk 'program' <file>
```

or

```
awk -f program.awk <file>
```

Each program has the structure

```
pattern {action}
```

The pattern is evaluated for each input line and if true, the statements in the action block are executed. Table 8.5 lists the most common patterns. Expressions and regular expressions can be combined with the logical operators `&&` (and), `||` (or), and `!` (not). Actions are specified through sequences of statements, some of which are listed in Table 8.6.

Statements are separated by newlines or semicolons; lines starting with `#` are comments. The most common input and output functions are listed in Table 8.7. One of these, `printf`, produces formatted output. Formatting is done via the format conversion commands listed in Table 8.8. Apart from `printf`, which prints to the screen (stdout), these are also recognized by `sprintf`, which prints to a string.

AWK has a number of built-in variables (Table 8.9). Of these, `NF`, the number of fields is particularly useful, as it allows traversal of all fields in a line, as in

```
for(i=1; i<=NF; i++)
    print $i
```

Table 8.5 Patterns in AWK

Pattern	Meaning
`BEGIN`	True before any input lines are processed
`END`	True after all input lines have been processed
Expression	Any expression in the AWK language
/regular expression/	True if matched
Pattern, pattern	Pattern range; true if in range

Table 8.6   AWK actions

Action	Meaning
delete *array element*	Delete specific entry from an array
exit	Terminate program
if(*expression*) *statement* [else *statement*]	Conditional execution
input–output statement	See Table 8.7
for(*expression*; *expression*; *expression*) *state-ment*	Repeat a fixed number of times
for(*variable* in *variable*) *statement*	Iterate over the keys of a hash
while(*expression*) *statement*	Execute while true
{*statement*}	Statements are grouped by curly brackets

Table 8.7   Input and output in AWK

Action	Meaning
close(*fileOrPipe*)	Close file or pipe
command │ getline	Pipe into getline
print	Print current line
printf *fmt*, *expr-list*	Print formatted output
system(*cmd*)	Send *cmd* to the shell for execution

Table 8.8   Formatting for printf and sprintf

Command	Meaning
%c	Character
%d	Decimal number
%e	Engineering convention, [-]d.dddddde[-+]dd
%f	Floating point number, [-]d.dddddd
%g	General: %d, %f, or %e, whichever is shorter
%s	String

Table 8.9   AWK built-in variables

Variable	Meaning
ARGC	Number of arguments on command line
ARGV	Array of arguments on command line, ARGV[0..ARGC-1]
FILENAME	Name of current input file
FS	Field separator
NF	Number of fields in current line
NR	Number of records (= lines)

Table 8.10 AWK string manipulation functions; s and t: strings, r: regular expression, i and n: integers

Function	Meaning
gsub(r, s, t)	Globally substitute r by s in t
index(s, t)	Return first starting position of t in s or 0 for no match
length(s)	Return length of s
match(s, r)	Return first starting position of r in s or 0 for no match
split(s, a, f)	Split s on f into fields saved in a, return number of fields; if f is omitted, use FS
sprintf(*fmt, expr-list*)	Return *expr-list* as a string formatted according to *fmt*
sub(r, s, t)	Like gsub, except that only first occurrence of r in t is replaced by s
substr(s, i, n)	Return n-char substring starting at i; if n is omitted, return suffix starting at i

Table 8.11 The arithmetic functions of AWK

Function	Meaning
cos(x)	$\cos(x)$
exp(x)	e^x
int(x)	Truncate x to integer
log(x)	Natural logarithm
rand()	Random number, $r, 0 \leq r < 1$
sin(x)	$\sin(x)$
sqrt(x)	$\sqrt{x}$
srand(x)	Seed the random number generator with x; otherwise, $x =$ current second

AWK is designed for manipulating strings and Table 8.10 lists its built-in string functions. Of these, sprintf has already been mentioned as allowing formatting.

AWK provides a selection of arithmetic functions, which are listed in Table 8.11.

Expressions are combined using the operators in Table 8.12. Perhaps, the most obscure—but in practice very useful—is string concatenation, which takes no explicit operator. For example, the code

```
a = "bio"          # assignment
b = "informatics" # assignment
c = a b            # concatenation & assignment
print c
```

would print bioinformatics.

To conclude this brief exposition of AWK, we give a few examples. Consider again our sample data in Fig. 8.2. Print the second column:

```
awk '{printf "%d\n", $2 }' data.dat
```

Table 8.12 AWK operators

Operators	Meaning
= += -= *= /= %= ^=	Assignment
? :	Conditional expression
||	OR
&&	AND
in	Key in hash
~ !~	Regular expression match and its negation
< <= > >= != ==	Comparisons
	String concatenation without explicit operator
+ -	Addition, subtraction
* /%	Multiplication, division, modulo
!	NOT
^	To the power of
++ -	Increment, decrement, can be used in prefix and postfix notation
$	field (column)

which replicates the command

```
cut -f 2 data.dat
```

However, in contrast to cut, AWK can also manipulate the input data. For instance, it can sum over the entries in the second column:

```
awk '{s += $2}END{printf "sum: %d\n", s}' data.dat
```

And their average is computed as

```
awk '{c++; s += $2}END{printf "avg: %g\n", s/c}' data.dat
```

AWK arrays behave like hash tables with strings or numbers as keys. In order to compute the occurrence of distinct numbers in the second column of the input data file, write

```
awk '{s[$2]++}END{for(a in s)printf "%d\t%d\n", a, s[a]}' data.dat
```

As a final example, consider

```
awk '/[25]/{print}' data.dat
```

as a replacement for our first example of a regular expression in grep,

```
grep [25] data.dat
```

Like the rest of UNIX, awk is described in its man pages. In addition, we have learned the language from the introduction by its authors, which is a model of clarity and usefulness [4].

8.2.9 Regular Expressions

The expression [25] we just used is an example of a regular expression, a notation for sets of strings; in this case, the set comprises the members "2" and "5". Another example is the dot (.), which matches any character, and hence refers to the set of all strings length one. As a rule, everything is text in UNIX, and as a consequence, regular expressions are used in many UNIX programs, not just the three examples we saw above, grep, sed, and AWK, but also in emacs and the shell. Knowing about regular expressions is thus very useful when working on UNIX systems.

There are three variants of regular expressions: regular, extended, and PERL (another programming language). In the following, we refer to the extended syntax, which is invoked by the -E switch in grep and the -r switch in sed. AWK regular expressions are by default of the extended kind. The sections "REGULAR EXPRESSIONS" in the man page for grep and "Regular Expressions" in the AWK man page contain more detail.

Character Classes

Character classes are written in square brackets. For example, [ab] matches either an a or a b. The complement of a character class is [^ab], which matches anything but a or b. Some character classes are used so frequently; there is a standardized notation for referring to them. We list six such classes in Table 8.13. To find out how, say, the character class "digit" works, try

```
sed 's/[[:digit:]]/x/g' data.dat
```

Quantifiers

There are four different types of quantifiers in regular expressions (Table 8.14). They are all known as *greedy*, which means they maximize the number of matches. For

Table 8.13 Regular expression character classes

Class	Meaning	Code
[:alpha:]	Letters	[A-Za-z]
[:digit:]	Digits	[0-9]
[:space:]	Whitespace characters	[t n r f v]
[:cntrl:]	Control characters	—
[:graph:]	Graphic characters	[^[:cntrl:]]
[:print:]	Printing characters	[[:graph]]

Table 8.14 Quantifiers in regular expressions

Number of matches (x)	Expression
$x \geq 0$	*
$x \geq 1$	+
$m \leq x \leq n$	{m,n}
$m \leq x$	{m, }

Table 8.15 Anchors in regular expressions

Expression	Explanation
\ b	Word boundary
^	Beginning of line (except when used inside a character class)
$	End of line

example, the expression . * would match the entire line of text rather than stopping at the beginning of the line upon encountering the first "match".

Anchors

Anchors allow a position within a string to be referenced, and Table 8.15 lists the three most important ones.

Backreferences

Assigning values to variables is one of the most important operations in traditional programming languages. In regular expressions, backreferences provide an analogous mechanism, of which there are two kinds, depending on where the referencing is done:

- Inside regular expressions: \n, where $1 \leq n \leq \infty$;
- Outside of regular expressions: $n, where $1 \leq n \leq \infty$.

For example, to substitute any pair of identical digits by just a single occurrence of that digit, sed could be used:

```
sed -E 's/([0-9])(\1)/\1/g' data.dat
```

Erratum to: Bioinformatics for Evolutionary Biologists

Erratum to:
B. Haubold and A. Börsch-Haubold,
***Bioinformatics for Evolutionary Biologists*,**
https://doi.org/10.1007/978-3-319-67395-0

The "Answers" and "Appendix: Unix Guide" have been reissued as last chapter of the book. Both the sections were previously listed as backmatter. No content within the chapter has been changed.

The updated online version of this book can be found at
https://doi.org/10.1007/978-3-319-67395-0

© Springer International Publishing AG 2018   E1
B. Haubold and A. Börsch-Haubold, *Bioinformatics*
for Evolutionary Biologists, https://doi.org/10.1007/978-3-319-67395-0_9

References

1. Abrahams, P.W., Larson, B.R.: UNIX for the Impatient, 2nd edn. Addison-Wesley (1996)
2. Adjeroh, D., Bell, T., Mukherjee, A.: The Burrows-Wheeler Transform:: Data Compression, Suffix Arrays, and Pattern Matching. Springer (2008)
3. Aho, A., Corasick, M.: Efficient string matching: an aid to bibliographic search. Commun. ACM **18**, 333–340 (1975)
4. Aho, A.V., Kernighan, B.W., Weinberger, P.J.: The AWK Programming Language. Addison-Wesley, Reading, MA (1988)
5. Aken, B.L., Achuthan, P., Akanni, W., Amode, M.R., Bernsdorff, F., Bhai, J., Billis, K., Carvalho-Silva, D., Cummins, C., Clapham, P., Gil, L., Girn, C.G., Gordon, L., Hourlier, T., Hunt, S.E., Janacek, S.H., Juettemann, T., Keenan, S., Laird, M.R., Lavidas, I., Maurel, T., McLaren, W., Moore, B., Murphy, D.N., Nag, R., Newman, V., Nuhn, M., Ong, C.K., Parker, A., Patricio, M., Riat, H.S., Sheppard, D., Sparrow, H., Taylor, K., Thormann, A., Vullo, A., Walts, B., Wilder, S.P., Zadissa, A., Kostadima, M., Martin, F.J., Muffato, M., Perry, E., Ruffier, M., Staines, D.M., Trevanion, S.J., Cunningham, F., Yates, A., Zerbino, D.R., Flicek, P.: Ensembl 2017. Nucl. Acids Res. **45**(D1), D635 (2017)
6. Altschul, S.F., Gish, W., Miller, W., Myers, E.W., Lipman, D.J.: Basic local alignment search tool. J. Mol. Biol. **215**, 403–410 (1990)
7. Altschul, S.F., Madden, T.L., Schäffer, A.A., Zhang, J., Zhang, Z., Miller, W., Lipman, D.: Gapped blast and psi-blast: a new generation of protein database search programs. Nucl. Acids Res. **25**, 3389–3402 (1997)
8. Baeza-Yates, R.A., Perleberg, C.H.: Fast and practical approximate string matching. Proceedings 3rd Symposium on Combinatorial Pattern Matching. Springer Lecture Notes in Computer Science, vol. 644, pp. 185–192. Springer, New York (1992)
9. Benjamini, Y., Hochberg, Y.: Controlling the false discovery rate: a practical and powerful approach to multiple testing. J. R. Stat. Soc. Ser. B **57**, 289–300 (1995)
10. Börsch-Haubold, A.G., Montero, I., Konrad, K., Haubold, B.: Genome-wide quantitative analysis of histone H3 lysine 4 trimethylation in wild house mouse liver: environmental change causes epigenetic plasticity. PlosOne **9**, e97,568 (2014)
11. Buck, L., Axel, R.: A novel multigene family may encode odorant receptors: a molecular basis for odor recognition. Cell **65**, 175–187 (1991)
12. Burrows, M., Wheeler, D.J.: A block-sorting lossless data compression algorithm. Technical Report 124, Digital Equipment Corporation, Palo Alto, California (1994)
13. Chang, J.T.: Recent common ancestors of all present-day individuals. Adv. Appl. Probab. **31**, 1002–1026 (1999)
14. Dayhoff, M.O., Schwartz, R.M., Orcutt, B.C.: A model of evolutionary change in proteins. In: M.O. Dayhoff (ed.) Atlas of Protein Sequence and Structure, vol. 5/suppl.3, pp. 345–352. National Biomedical Research Foundation, Washington DC (1978)

© Springer International Publishing AG 2017

B. Haubold and A. Börsch-Haubold, *Bioinformatics for Evolutionary Biologists*, https://doi.org/10.1007/978-3-319-67395-0

15. Delcher, A.L., Kasti, S., Fleischmann, R.D., Peterson, J., White, O., Salzberg, S.L.: Alignment of whole genomes. Nucl. Acids Res. **27**, 2369–2376 (1999)
16. Farris, J.S.: Estimating phylogenetic trees from distance matrices. Am. Nat. **106**, 645–668 (1972)
17. Felsenstein, J.: Inferring Phylogenies. Sinauer, Sunderland (2004)
18. Gau, J., Watabe, H., Aotsuka, T., Pang, J., Zhang, Y.: Molecular phylogeny of the Drosophila obscura species group, with emphasis on the old world species. BMC Molecular. Evol. Biol. **7**, 87 (2007)
19. Gibbs, A.J., McIntyre, G.A.: The diagram, a method for comparing sequences; its use with amino acid and nucleotide sequences. Eur. J. Biochem. **16**, 1–11 (1970)
20. Gupta, S.K., Kececioglu, J.D., Schäffer, A.A.: Improving the practical space and time efficiency of the shortest-path approach to sum-of-pairs multiple sequence alignment. J. Comput. Biol. **2**, 459–472 (1995)
21. Gusfield, D.: Algorithms on Strings, Trees, and Sequences: Computer Science and Computational Biology. Cambridge University Press, Cambridge (1997)
22. Haig, D., Hurst, L.D.: A quantitative measure of error minimization in the genetic code. J. Mol. Evol. **33**, 412–417 (1991)
23. Haubold, B., Klötzl, F., Pfaffelhuber, P.: andi: Fast and accurate estimation of evolutionary distances between closely related genomes. Bioinformatics **31**, 1169–75 (2015)
24. Haubold, B., Wiehe, T.: Introduction to Computational Biology: An Evolutionary Approach. Birkhäuser, Basel (2006)
25. Hudson, R.R.: Generating samples under a Wright-Fisher neutral model of genetic variation. Bioinformatics **18**, 337–338 (2002)
26. Jukes, T.H., Cantor, C.R.: Evolution of protein molecules. In: Munro, H.N. (ed.) Mammalian Protein Metabolism, vol. 3, pp. 21–132. Academic Press, New York (1969)
27. Kasai, T., Lee, G., Arimura, H., Arikawa, S., Park, K.: Linear-time longest-common-prefix computation in suffix arrays and its applications. LNCS **2089**, 181–192 (2001)
28. Knuth, D.E.: The TEXbook. Addison-Wesley, Reading, Massachusetts (1994)
29. Kurtz, S., Phillippy, A., Delcher, A., Smoot, M., Shumway, M., Antonescu, C., Salzberg, S.: Versatile and open software for comparing large genomes. Genome Biol. **5**(2), R12 (2004)
30. Lamport, L.: A Document Preparation System: LATEX, 2nd edn. Addison-Wesley, Boston (1994)
31. Lee, H., Popodi, E., Tang, H., Foster, P.L.: Rate and molecular spectrum of spontaneous mutations in the bacterium Escherichia coli as determined by whole-genome sequencing. Proc. Natl. Acad. Sci. USA **109**, E2774–83 (2012)
32. Lempel, A., Ziv, J.: On the complexity of finite sequences. IEE Trans. Inf. Theory **IT-22**, 75–81 (1976)
33. Li, H., Durbin, R.: Fast and accurate short read alignment with burrows-wheeler transform. Bioinformatics **25**, 1754–1760 (2009)
34. Lipman, D.J., Altschul, S.F., Kececioglu, J.D.: A tool for multiple sequence alignment. Proc. Natl. Acad. Sci. USA **86**, 4412–4415 (1989)
35. Lynch, M.: The Origins of Genome Architecture. Sinauer, Sunderland (2007)
36. Manber, U., Myers, E.W.: Suffix arrays: a new method for on-line string searches. SIAM J. Comput. **22**, 935–948 (1993)
37. Needleman, S.B., Wunsch, C.D.: A general method applicable to the search for similarities in the amino acid sequence of two proteins. J. Mol. Biol. **48**, 443–453 (1970)
38. Ohlebusch, E.: Bioinformatics Algorithms: Sequence Analysis, Genome Rearrangements, and Phylogenetic Reconstruction. Enno Ohlebusch, Ulm (2013)
39. Plank, L.D., Harvey, J.D.: Generation time statistics of Escherichia coli B measured by synchronous culture techniques. J. Gen. Microbiol. **115**, 69–77 (1979)
40. Rice, W.R.: Analyzing tables of statistical tests. Evolution **43**, 223–225 (1989)
41. Rohde, D.L.T., Olson, S., Chang, J.T.: Modelling the recent common ancestry of all living humans. Nature **431**, 562–566 (2004)

42. Saitou, N., Nei, M.: The neighbor-joining method: a new method for reconstructing phylgenetic trees. Mol. Biol. Evol. **4**, 406–425 (1987)
43. Sanger, F., Coulson, A.R., Hong, G.F., Hill, D.F., Petersen, G.B.: Nucleotide sequence of bacteriophage λ DNA. J. Mol. Biol. **162**, 729–773 (1982)
44. Smith, T.F., Waterman, M.S.: Identification of common molecular subsequences. J. Mol. Biol. **147**, 195–197 (1981)
45. Student: The probable error of a mean. Biometrica **6**, 1–25 (1908)
46. Swofford, D.L., Olsen, G.J., Waddell, P.J., Hillis, D.M.: Phylogenetic inference. In: Hillis, D.M., Craig, M., Marble, B.K. (eds.) Molecular Systematics, 2nd edn, pp. 407–514. Sinauer, Sunderland (1996)
47. The Gene Ontology consortium: Gene ontology: tool for the unification of biology. Nature Genetics **25**, 25–29 (2000)
48. Ukkonen, E.: On-line construction of suffix-trees. Algorithmica **14**, 249–260 (1995)
49. Wakeley, J.: Coalescent Theory: An Introduction. Roberts & Company, Colorado (2009)
50. Watterson, G.A.: On the number of segregating sites in genetical models without recombination. Theor. Popul. Biol. **7**, 256–276 (1975)
51. Wikipedia: Observable universe (2016). http://en.wikipedia.org/wiki/Observable_universe

Index

© Springer International Publishing AG 2017
B. Haubold and A. Börsch-Haubold, *Bioinformatics for Evolutionary Biologists*, https://doi.org/10.1007/978-3-319-67395-0

Printed in the United States
By Bookmasters